A Guide to Understanding
Basic Organic Reactions

R. C. Whitfield MA BSc PhD ARIC

Lecturer in Education
Cambridge University
formerly Assistant Master, Cheltenham College

Longmans

Longmans, Green & Co Ltd
London & Harlow
Associated companies, branches and representatives
throughout the world

© R. C. Whitfield 1966
First Published 1966
Third impression 1969

SBN 582 32143 3

Printed in Hong Kong by Peninsula Press Ltd.

Preface

The traditional method of teaching elementary organic chemistry has largely involved the systematic study of the preparation and properties of the various classes of organic compounds. This method, which is still the most common initial approach to the subject, places before the pupil some of the wealth of factual material which organic chemists have known for many years. All too frequently however, especially in schools, little attempt is made to introduce the underlying physico-chemical principles of organic reactions which are so vital to the modern chemist. Some mention of the forces which are the impetus for reactions to take place, and the intermediate stages whereby reactants yield products, must be included in any satisfactory introduction to organic chemistry. The average pupil is otherwise left somewhat bewildered by the vast number of apparently unrelated reactions which he has to commit to memory, and the important question ' *Why* do A and B react to give C and D ?', is left unanswered.

The aim of this monograph is to explain some basic organic reactions, in order that the school and technical college pupil may understand something of the way in which organic molecules react. The reactions which have been selected are drawn from varying sectors of organic chemistry, and almost all will have been covered during Advanced Level, Scholarship Level and National Certificate courses.

After a brief introduction, the reactions are divided into four main classes: substitution, addition, elimination, and the less familiar rearrangement reactions. Several reactions appear in each class, with as little overlapping of mechanistic type as possible. The application of several of the measurements of physical chemistry in the elucidation of reaction mechanisms is illustrated in many of the reactions discussed. New reaction principles are introduced gradually as required in an attempt to link basic theory with real reactions, this being the most common classroom teaching method. Exercises, selected for the reader to test his understanding and ability to apply the text and predict reaction paths, are given at the end of each section. By use of these exercises, a significant number of different reactions which have similar mechanisms will also be covered. It is therefore hoped that the student with but little previous knowledge of organic chemistry will be able to work through the text on his own. A working knowledge of atomic theory, molecular structure, elementary reaction kinetics and isomerism is, however, assumed. The names of many of the various electronic effects in organic chemistry have been deliberately omitted, because I believe that these can be more confusing than helpful to the majority of pupils at this level.

The book is intended to stimulate those pupils who have more than the present 'A'-level requirements of the examining boards as their ultimate horizon, and who wish to find some logic in organic chemistry, as opposed to a mere list of reactions to be learned. It is my firm belief and personal experience that once the pupil understands the basic principles of organic reactions, then not only will his interest in organic chemistry be stimulated, but also, his task of memorizing the large number of reactions will be greatly simplified.

I wish to thank Dr D. I. Davies, Lecturer in Organic Chemistry at King's College, London, who kindly read the manuscript during its preparation and made many helpful comments. I am also greatly indebted to several of the authors whose names appear in the bibliography, notably T. A. Geissman and Peter Sykes.

Cheltenham College   Richard Whitfield
May 1965

A Note to the Reader

This book will make you think, and you should read it carefully and thoughtfully. There may be several concepts which are new to you, and you should make sure that you understand these as they are introduced. You are recommended to answer the questions provided at the end of each section in order to test your comprehension of the text as you proceed; answers are provided. The mechanistic method of examining organic reactions inevitably carries with it some terms with which you may be unfamiliar; once you have mastered these early on in the book, you should be able to proceed with your understanding of organic reactions much more rapidly. A glossary is provided for quick reference.

Contents

1 Introduction

During the study of chemical reactions, and organic reactions in particular, two factors must be continually borne in mind. Firstly, that atoms and molecules are tangible objects, each possessing both shape and size. Most atoms have radii between about 0·4 and 1·5 Angström units (1 Angström unit, Å, = 10^{-8} cm), and the lengths of chemical bonds are usually in the range 0·7 to 3·0Å. The heavier atoms and molecules are larger than the lighter ones; for example, the relative volumes of hydrogen, carbon, chlorine and iodine atoms are approximately 1:10:20:50. The shapes of molecules are largely determined by the number and type of electrons in the individual atoms which are available for bond formation, and by the character of the bonds which are produced. Carbon compounds contain predominantly covalent bonds. These bonds are directional in character because the electron distribution is concentrated between the combined atoms. The shapes of organic molecules — the way they appear in three dimensions — are therefore a vital factor in a study of organic reactions.

Secondly, chemical reaction of one atom, molecule, or ion with another always involves a redistribution of electrons; changes in the distribution of the electrons round the atomic nucleii always accompany chemical change. Consequently, if we are to be able to understand the mechanics of chemical reactions, we must appreciate how and why electrons move and redistribute themselves during the course of a reaction, to yield products differing in chemical structure from the starting materials.

If then, we consider (1) the shapes and sizes of atoms and molecules (that is, the consideration of 'stereochemistry') and (2) the movement and redistribution of electrons during reactions, we should be in some position to predict what might happen if we added one chemical compound to another under a given set of conditions. This is surely the ultimate goal of chemical science — to be able to predict the paths of the interconversion of matter. Several of the exercises provided in the book will test your ability to predict organic reaction paths on the basis of principles which will be encountered.

1.1 The four types of organic reaction

It is most convenient that the numerous reactions of the various classes of organic compounds can all be classified into four basic reaction types:

1. Substitution reactions
2. Addition reactions
3. Elimination reactions
4. Rearrangement reactions

Substitution reactions are those in which an atom or group of atoms in a molecule is displaced by a different atom or group of atoms; one group leaves the molecule and another takes its place, for example:

$$CH_3-X+Y \xrightarrow{\text{substitution}} CH_3-Y+X$$

Addition reactions are those in which atoms or groups of atoms are added to a molecule; no part of the original molecule is lost, there is simply a net gain of the reagent atoms in the product molecule. This type of reaction only occurs where there is a centre of unsaturation in a molecule, that is, when there is a multiple bond between atoms, as in, for example, olefins $\left(\diagup C = C \diagdown\right)$, acetylenes ($-C{\equiv}C-$), cyanides ($-C{\equiv}N$), and aldehydes and ketones $\left(\diagup C = O\right)$, for example:

$$CH_2{=}CH_2+XY \xrightarrow{\text{addition}} \underset{X\ \ \ \ Y}{CH_2-CH_2}$$

Elimination reactions are essentially the reversal of addition reactions. Atoms or groups of atoms are completely removed from a molecule and are not replaced by any other group. A new multiple linkage therefore, always appears in the product. Most commonly, loss of atoms or groups occurs from adjacent carbon atoms to yield an olefin, for example:

$$\underset{X\ \ \ \ Y}{CH_2-CH_2} \xrightarrow{\text{elimination}} CH_2{=}CH_2+XY$$

Rearrangement reactions either involve (1) the migration of a functional group to another position in the molecule, an olefinic bond often being involved, for example:

$$\underset{X}{CH_3-CH-CH{=}CH_2} \xrightarrow[\text{migration}]{\text{functional group}} CH_3-CH{=}CH-CH_2-X$$

or (2) the actual rearrangement of the basic carbon skeleton of a molecule, for example:

$$\underset{CH_3}{\overset{CH_3}{CH_3-C-CH_2-X}} \xrightarrow[\text{rearrangement}]{\text{carbon skeleton}} \underset{X}{\overset{CH_3}{CH_3-C-CH_2-CH_3}}$$

The examples of rearrangement reactions discussed later show that addition, elimination and substitution stages often occur subsequent to the rearrangement step.

Reactions in which a larger molecule is produced by the coupling of two smaller molecules, often with the formation of a simple molecule such as water, have been traditionally termed 'condensation reactions'. These do *not* form a fifth class of organic reaction, since they are primarily addition reactions which are often followed by an elimination stage, for example:

$$
\underset{O}{\overset{H}{\underset{\|}{CH_3-\overset{|}{C}}}} + H_2N-NH_2 \xrightarrow[\text{across C=O}]{\text{addition}} \underset{OH}{\overset{H}{\underset{|}{CH_3-\overset{|}{C}-NH-NH_2}}}
$$

Intermediate addition
compound — not isolated

$$\downarrow \begin{array}{l} \text{elimination} \\ \text{of H—OH} \end{array}$$

$$CH_3-CH\!\!=\!\!N-NH_2 + H_2O$$

Acetaldehyde hydrazone

1.2 Bond fission and formation

During the course of any chemical reaction, a chemical bond between two atoms must be ruptured. An electron pair bond between two atoms A and B can break in three possible ways depending upon the relative electronegativities, or 'electron attracting powers', of A and B.

(1) $A: \mid B \rightarrow A:^{\ominus} + B^{\oplus}$ (A more electronegative than B)

(2) $A \mid :B \rightarrow A^{\oplus} + :B^{\ominus}$ (B more electronegative than A)

(3) $A \,{\cdot}\!:^{\cdot} B \rightarrow A\cdot + \cdot B$ (A and B usually of similar electronegativity)

The first two modes of bond fission result in the formation of ions. This process is termed 'heterolytic fission', and it results in unequal sharing of the electron pair by the ionic entities produced. The third mode of fission results in the formation of very reactive 'free radicals'. These are atoms or groups of atoms with a single odd unpaired electron. This process is termed 'homolytic fission', and equal sharing of the electron pair by the resulting entities takes place.

As we shall see, processes (1) and (2) are the more common in organic reactions, so that although organic compounds possess predominantly covalent structures, ions feature to a large extent in many of their reactions.

Bond fission may precede or be simultaneous with the formation of the new chemical bond in a reaction. For example, an ion $:C^\ominus$ may react in substitution reactions with the molecule AB of type (1) above in either of the following ways:

(a) $A:|B \xrightarrow[\text{bond fission}]{\text{first step}} A:^\ominus + B^\oplus$

Then $B^\oplus + :C^\ominus \xrightarrow[\text{bond formation}]{\text{second step}} B:C$

(b) $A:|B$ $\xrightarrow[\text{formation simultaneous}]{\text{bond fission and}} A:^\ominus + B:C$
 $\curvearrowleft :C^\ominus$

($:C^\ominus$ is the reagent; $A:B$ is the substrate; $B:C$ is the product.)

In the first case the ion $B^\oplus$ is an *intermediate* species with probably a short lifetime. In case (b) it has a zero lifetime, the reaction proceeding through an intermediate 'transition state' which can be represented by

$$\overset{\delta-}{A:} ---\leftarrow-- B ---\leftarrow-- \overset{\delta-}{:C}$$

bond   bond
breaking   making

The formation of transient intermediate moieties is a common feature of organic reactions.

If *both* electrons of the new electron pair bond formed in the product are provided by one of the reacting species, the reaction is said to proceed via an *ionic* mechanism. If each reacting centre contributes *one* electron to the new bond, the reaction is said to proceed via a *free radical* mechanism.

1.3 Classification of reagents

The species $A:^\ominus$ and $:C^\ominus$ each possess a lone pair of electrons. They are therefore capable of acting as bases, the extended concept of a base being a 'proton acceptor' (Lowry-Brönsted) or 'an electron pair donor' (Lewis). They are said to be *nucleophilic* ('nucleus loving' from the Greek *filia* = to like or to love) entities.

The species $B^\oplus$ is on the other hand electron deficient, and it is capable of accepting an electron pair. It is therefore a Lewis acid and is said to be an *electrophilic* ('electron loving') entity.

There is a natural attraction between electrophilic and nucleophilic entities, for the electron deficiency of the one is capable of being satisfied by the electron availability in the other. In reactions which occur via an ionic mechanism, the new electron pair bond is formed by the union of a nucleophile (nucleophilic entity) with an electrophile (electrophilic entity), the electrons forming the bond being provided by the nucleophile, for example:

$$B^{\oplus} \quad + \quad :C^{\ominus} \rightarrow \quad B:C$$

electrophile nucleophile product

In the general substitution reaction we have been discussing, namely

$$A:B \quad +:C^{\ominus} \rightarrow A:^{\ominus}+ \quad B:C$$

substrate reagent product

the reagent entity $:C^{\ominus}$, which attacks atom B by either the one stage or the two stage process to form the new bond $B:C$, is a nucleophile. The total process, conventionally named after the reagent, is termed a nucleophilic substitution reaction.

The concept of electrophilic and nucleophilic entities will be occurring frequently in our study of the mechanics of organic reactions. The summary table gives their characteristics.

SUMMARY TABLE

Nucleophilic entities (Nucleophiles)	Electrophilic entities (Electrophiles)
Electron rich	Electron deficient
Provide an electron pair	Accept an electron pair
Lewis base	Lewis acid
Possess an unshared pair of electrons which are not too strongly held to the atomic nucleus. (Usually atoms in groups V and VI of the Periodic Table.)	Possess an empty orbital to receive the electron pair from the nucleophile
Are able to increase their covalency by one unit	Are able to form an extra or alternative bond with the nucleophile
Often anions	Usually cations
Examples: H_2O, ROH, $OH^{\ominus}$, $RO^{\ominus}$, $Br^{\ominus}$, NH_3, RNH_2, $CN^{\ominus}$.	Examples: $H^{\oplus}$, $Br^{\oplus}$, $R-\overset{\oplus}{N}\equiv N$, $R_3C^{\oplus}$, $NO_2^{\oplus}$.

6

Questions

1 Complete the sentences:

   (a) The heterolytic fission of an electron pair bond gives rise to ....

   (b) The water molecule is a nucleophilic entity because it can .... an electron pair for bonding to an .... entity. Water is also a Lewis ....

2 By means of a clear electronic diagram, indicate how a chlorine molecule would form two chlorine radicals. Are these radicals electrically neutral?

2 Substitution Reactions

2.1 The replacement of hydroxyl by halogen

The reaction

$$CH_3CH_2OH + HBr \rightarrow CH_3CH_2Br + H_2O$$

is an example of the general reaction of alcohols with concentrated solutions of halogen acids.

Alcohols can behave as weak bases. Under suitably acidic conditions they will coordinate a proton by lone pair donation, for example:

$$CH_3CH_2-\overset{..}{\underset{\diagdown}{O}}: + H^{\oplus} \rightarrow \left[CH_3CH_2 \diagdown \overset{..}{\underset{|}{O}} \diagup H \right]^{\oplus}$$
$$\quad\quad\quad H \quad\quad\quad\quad\quad\quad H$$

Compare

$$H-\overset{..}{\underset{\diagdown}{O}}: + H^{\oplus} \rightarrow \left[H \diagdown \overset{..}{\underset{|}{O}} \diagup H \right]^{\oplus}$$
$$\quad\quad H \quad\quad\quad\quad\quad H$$

Here, the extended concept of a base as a 'proton acceptor' (Lowry-Brönsted) or electron donor (Lewis) is being applied. Alcohols are therefore nucleophilic reagents.

Protonation of alcohols is usually the first stage in their reaction with common inorganic strong acids. In our example, when ethanol is treated with concentrated aqueous hydrobromic acid, the solution will therefore contain $H_3O^{\oplus}$, $Br^{\ominus}$, and $CH_3CH_2\overset{\oplus}{O}H_2$ ions. The last of these three ions, namely protonated ethanol, will contain a much weakened C—O bond since the positive charge, formally located on the oxygen atom, will attract electrons from the C—O bond towards the oxygen nucleus. This will in turn render the first carbon atom electron deficient and it will therefore be able to accept electrons from any available nucleophilic entity.

Available nucleophiles in solution are $Br^{\ominus}$ ions and neutral ethanol or water molecules. These are the species which can potentially attack the electron deficient carbon atom. Bonding of any one of these species to carbon cannot however take place until the C—O bond actually breaks, as carbon does not possess five available orbitals. Hence, if we indicate the shift of a *pair* of electrons by curved arrows, then the reaction with bromide ion as nucleophile can be pictured

$$\left[:\overset{..}{\underset{..}{Br}} : \right]^{\ominus} \quad \overset{CH_3}{\underset{CH_2 \odot OH_2}{|}} \rightarrow :\overset{..}{\underset{..}{Br}} : \overset{CH_3}{\underset{CH_2}{|}} + H_2O$$

ethyl
bromide

or, more simply

$$Br:^{\ominus} \longrightarrow \overset{\overset{\displaystyle CH_3}{|}}{CH_2} \overset{\oplus}{\longrightarrow} OH_2 \rightarrow CH_2Br + H_2O$$

The reason why the shift of a *pair* of electrons neutralizes the *single* positive charge formally located on the oxygen atom is that oxygen already effectively 'owns' one of the electrons of the electron pair bond, or more precisely, has a half-share of each of the two electrons in the covalent bond. Hence if the electron pair shifts on to oxygen there is in fact a net gain of only *one* electron. Similarly, if bromide ion donates an electron pair for covalent bond formation to carbon, there is only a *net* loss of one electron by bromide, which neutralizes its unit negative charge. *It is vital that this point is fully understood.*

The formation of the new bond between bromine and carbon may occur at the same time as the carbon–oxygen bond breaks, as the above written mechanism suggests. Alternatively, rupture of the carbon–oxygen bond may precede carbon–bromine bond formation.

$$CH_3CH_2 \overset{\oplus}{\underset{}{OH_2}} \xrightarrow[\text{out}]{\text{water}} H_2O + \overset{(1)}{CH_3CH_2} \quad \overset{Br:^{\ominus}}{\underset{\text{Intermediate}}{\xrightarrow[\text{ion in}]{\text{bromide}}}} CH_3CH_2Br$$
carbonium ion

The factors affecting, and the peripheral consequences of these alternative mechanisms are too complex to discuss here. The important feature to be grasped about the reaction, is that bromide ion displaces water from protonated ethanol by nucleophilic attack on carbon, proceeding through some intermediate or 'transition' state.

The net reaction therefore of an alcohol R—OH with a concentrated aqueous solution of hydrobromic acid (which can be generated *in situ* from sodium bromide and concentrated sulphuric acid) is

$$HBr + R{-}OH \overset{\text{heat}}{\rightarrow} R{-}Br + H_2O$$
$$\text{alcohol} \qquad \text{alkyl}$$
$$\text{bromide}$$

that is, hydroxyl is replaced by halogen.

These reaction conditions yield a high concentration of bromide ions, and this ion is therefore more likely to attack the electron deficient carbon atom than any other nucleophile present to yield the alkyl bromide.

PRINCIPLE INTRODUCED IN THIS SECTION

The conventional use of curved arrows to depict electron pair shifts.

Question

What is likely to be the major product from the reaction of ethanol with aqueous hydrobromic acid if ethanol, rather than bromide ions, are present in excess in the reaction mixture? Draw out the path of this reaction using curved arrows to depict electron pair shifts.

2.2 The Williamson synthesis of ethers

The reactive alkali metals, sodium and potassium, dissolve readily in alcohols to yield the metal alkoxide, hydrogen being liberated, for example:

$$2CH_3CH_2OH + 2Na \rightarrow 2CH_3CH_2O^\ominus Na^\oplus + H_2 \uparrow$$
sodium ethoxide
(a metal alkoxide)

Metal alkoxides are strong organic bases and very effective nucleophiles. They are usually used as a solution in the corresponding alcohol and have a wide application in organic chemistry. Their use in the Williamson ether synthesis involves reaction with primary alkyl halides, for example:

$$CH_3CH_2O^\ominus Na^\oplus + CH_3CH_2Br \rightarrow CH_3CH_2OCH_2CH_3 + Na^\oplus Br^\ominus$$
ethyl bromide diethyl ether
(an alkyl halide)

Before we can gain insight into this reaction, a new principle must be discussed.

ELECTRON ATTRACTION BY HALOGEN: INDUCTIVE EFFECT

The halogen elements possess electronegativity coefficients ranging from 4·0 to 2·5 on the Pauling scale. (See table)

Table of electronegativity values

			H 2·1			
Li	Be	B	C	N	O	F
1·0	1·5	2·0	2·5	3·0	3·5	4·0
Na	Mg	Al	Si	P	S	Cl
0·9	1·2	1·5	1·8	2·1	2·5	3·0
						Br
						2·8
						I
						2·5

The value for carbon is 2·5, and therefore the alkyl halides (alkyl iodides possibly excepted) possess carbon–halogen bonds with a certain degree of ionic character; these bonds are not wholly covalent. The halogen atom, X, acquires a slightly greater than half share in the pair of electrons which make up the bond, due to its greater electron attracting power over carbon. This can be depicted in terms of the relative closeness of the electron pair to a particular atom, or the relative density of the electron pair cloud or by the allocation of fractional formal charges:

An '*inductive pull*' of electrons towards halogen exists, and the physical evidence for this comes from two main sources; acid strengths, and dipole moments.

Acid strengths. The relative strengths of acids can be judged from their pK_a values, where pK_a signifies $-\log_{10}$ (acid dissociation constant). A low pK_a value signifies a strong acid, that is, one which is ionized to a proton and an anion to a significant extent. Organic carboxylic acids RCOOH are partially ionized.

$$RCOOH \rightleftharpoons RCOO^{\ominus} + H^{\oplus} \quad \text{(Proton as } H_3O^{\oplus} \text{ in aqueous solution)}$$

Variation of the group R gives rise to acids of different strengths which are attributed to the bond properties inherent in the group R.

pK_a *values of some carboxylic acids*

Name	Formula	pK_a
Formic acid	$H \cdot COOH$	3·8
Acetic acid	$CH_3 \cdot COOH$	4·8
Propionic acid	$CH_3 \cdot CH_2 \cdot COOH$	4·9
Chloroacetic acid	$Cl \cdot CH_2 \cdot COOH$	2·8
Dichloroacetic acid	$Cl_2CH \cdot COOH$	1·3
Trichloroacetic acid	$Cl_3 \cdot C \cdot COOH$	0·1
Bromoacetic acid	$Br \cdot CH_2 \cdot COOH$	2·9
Iodoacetic acid	$I \cdot CH_2 \cdot COOH$	3·0
α-Chloropropionic acid	$CH_3 \cdot CHCl \cdot COOH$	2·8
β-Chloropropionic acid	$Cl \cdot CH_2 \cdot CH_2 \cdot COOH$	4·1

The table shows that halogen substituted acetic acids are significantly stronger than acetic acid and that increasing halogenation results in increasing acidity. In fact, in 0·3 M aqueous solution, the percentage ionization of acetic and trichloroacetic acids is 2 per cent and 90 per cent respectively. The carboxyl group, —COOH, is common to all the acids, and the varying acid strengths must therefore be due to the atoms or groups attached to the carbon atom adjacent to the carboxyl group (the α-carbon atom). The halogen atoms inductively withdraw electrons from the carbon atom to which they are bound, and this electron withdrawal is transmitted through the α-carbon atom away from the carboxyl group. Thus, for chloroacetic acid we have

$$Cl—\leftarrow CH_2—\leftarrow \overset{\displaystyle \overset{O}{\|}}{C}—\leftarrow O—\leftarrow H$$

$$\underset{\text{the α-carbon atom}}{\uparrow}$$

where the arrows depict the direction of electron displacement induced by the electronegativity of the chlorine atom. This general 'electron suck' towards halogen results in a weakening of the O—H bond of the carboxyl group, and deprotonation and ionization of the acid is favoured.

The more electronegative the halogen, the more effective the electron displacement and the stronger the acid. Thus bromoacetic and iodoacetic acids are weaker than chloroacetic acid. Despite the equality of electronegativity values for carbon and iodine, it should be noted that iodo-acetic acid is still a stronger acid than acetic acid, indicating that the C—I bond must also tend to inductively withdraw electrons from the carboxyl group.

Greater substitution of halogen at the α-carbon atom also gives rise to a more effective electron displacement and a corresponding weakening of the O—H bond. This accounts for the increasing acidity of the mono, di- and trichloroacetic acids.

The inductive pull of electrons away from the carboxyl group however diminishes markedly as the halogen atom is moved along the carbon chain away from the carboxyl group. Even when halogen resides on the β-carbon atom, as in β-chloropropionic acid, an acidity comparable with an unsubstituted acid results.

$$Cl—\leftarrow CH_2—\leftarrow CH_2—\overset{\displaystyle \overset{O}{\|}}{C}—O—H$$

$$\underset{\substack{\text{the} \\ \text{β-carbon} \\ \text{atom}}}{\uparrow} \quad \underset{\substack{\text{the} \\ \text{α-carbon} \\ \text{atom}}}{\uparrow}$$

The inductive pull of electrons is not efficiently transmitted along more than one carbon–carbon bond.

Another deduction which can be made from the foregoing table of pK_a values is that alkyl groups are effectively electron donors. Acetic and propionic acids are weaker than formic acid, indicating that ionization to a proton and a carboxylate anion becomes less easy if the carboxyl group is attached to alkyl rather than hydrogen. The alkyl group must therefore tend to push electrons along on to the carboxyl group thus strengthening the O—H bond and discouraging ionization.

$$CH_3 \rightarrow \overset{\overset{\textstyle O}{\|}}{C} \rightarrow O \rightarrow H$$

The potential acidic hydrogen atom is more tightly bound in the 'electron sea' of the molecule in these cases. We shall be mentioning this phenomenon, which is the reverse of the effect of halogens, later in the book (page 58).

It should be emphasized that the electron displacements referred to in this section only involve shifts within the individual covalent linkages and not displacements out of one bond into another.

Dipole moments. A molecular dipole moment is a measure of the electrical asymmetry of a molecule. The asymmetric positioning of the electron cloud in a chemical bond results in a separation of charge within the molecule, and there is then a tendency for the molecule to line up along the direction of an electrostatic field. For example, in a molecule XY in which atom Y was more electronegative than X we would have:

$$\ominus \quad \overset{\delta+ \quad \delta-}{X—Y} \quad \oplus$$

$$\leftarrow \text{ electric field } \rightarrow$$

The value of the dipole moment is the product of the charge size and the distance of separation in the molecule, and this is experimentally measurable to a high degree of accuracy. The normal unit of dipole moment is the Debye unit (D); this is equal to 10^{-18} e.s.u. cm. Two electrons (each of charge 4.8×10^{-10} e.s.u.) separated by a distance of 1 Å would have a dipole moment of 4.8 D. For polyatomic molecules the molecular dipole moment is the vector sum of the individual bond moments in the molecule. The dipole moment of methane is zero as there is a completely symmetrical electron distribution in this molecule. The alkyl halides however have dipole moments ranging from about 1.3 to 2.0 D, thus confirming the separation of charge in these molecules resulting from electron attraction by halogen. Carbon tetrachloride, like methane, has no dipole moment. Here the tetrahedrally disposed individual C—Cl bond moments cancel, vectorially adding up to zero.

THE REACTION SEQUENCE

The two major driving forces for the reaction under consideration are

1 the strong nucleophilic power of the alkoxide anion, and
2 the ability of halogen to form a stable anion $X^{\ominus}$.

The reaction proceeds via nucleophilic attack by alkoxide ion on the carbon atom bearing the halogen atom, the halogen atom being expelled as its anion.

$$CH_3-CH_2\ddot{\underset{..}{O}}:^{\ominus} \longrightarrow \underset{\delta+\quad\delta-}{CH_2-Br} \rightarrow CH_3-CH_2-\ddot{\underset{..}{O}}-CH_2-CH_3 + Br^{\ominus}$$

with CH_3 on the carbon bearing Br.

If the halogen is fluorine, chlorine, or bromine, then there is on the basis of electronegativity values an additional impetus for the reaction to take place. The carbon atom bearing the halogen is slightly electron deficient (has $\delta+$ charge) before reaction takes place; its affinity for the electron rich alkoxide ion will therefore initially be greater than in the case of an alkyl iodide. However, there is another factor which has some bearing on the ease of reaction. As the halogen atoms become larger with increasing atomic weight, the carbon–halogen bond lengths in the alkyl halides become longer. Furthermore, the bond energy, which is a measure of the energy required to break the bond, becomes less.

Bond energies of carbon–halogen bonds (in kcal/mole)

C—F	C—Cl	C—Br	C—I
105	79	66	57

Increasing length of bond
—————————————→
Decreasing strength of bond

This is chiefly because the bonding electrons in larger atoms are less under the control of the atomic nucleus; they are more accessible; their charge cloud is more readily distorted. Thus, when an electron rich nucleophilic entity approaches the carbon–iodine bond of an alkyl iodide, an electron repulsion takes place along the bond. This repulsion is towards iodine because its electron cloud is easily distorted. Thus, although the alkyl iodides do not possess an inherent inductive pull of electrons towards iodine, an analogous drift of charge does take place as the nucleophile approaches.

The iodides therefore take part in the Williamson ether synthesis very satisfactorily by a similar mechanism, for example:

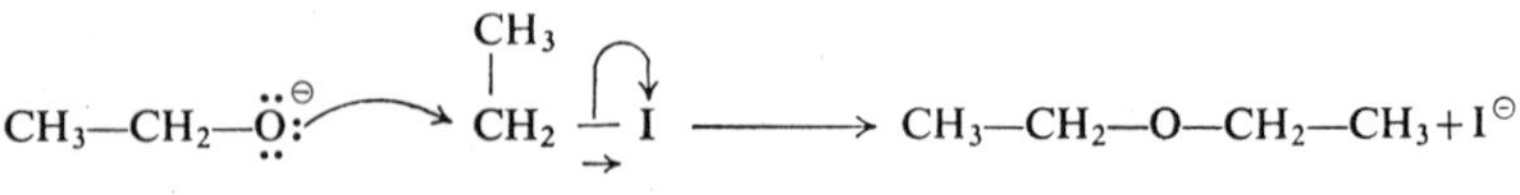

$$CH_3—CH_2—\overset{..}{\underset{..}{O}}{:}^{\ominus} \longrightarrow \overset{CH_3}{\underset{}{\underset{|}{CH_2}}} \overset{}{\underset{\rightarrow}{—}} I \longrightarrow CH_3—CH_2—O—CH_2—CH_3 + I^{\ominus}$$

Drift of electrons
as nucleophile
approaches

Secondary and tertiary alkyl halides of the type

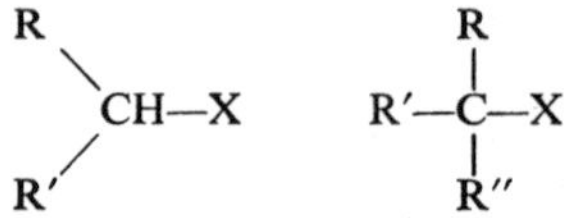

$$\underset{R'}{\overset{R}{\diagdown}} CH—X \qquad R'—\overset{R}{\underset{R''}{\overset{|}{C}}}—X$$

Secondary   Tertiary

do not participate so successfully in this reaction, largely because the bulky alkyl groups R etc. hinder the approach of the alkoxide ion to the carbon atom where displacement occurs. The attacking entity is prevented from reaching the reaction site in these cases by spatial congestion.

PRINCIPLE INTRODUCED IN THIS SECTION

The inductive effect. This is the unequal sharing of bonding electrons by the atoms which they join. It is the result of differences in electronegativity of the linked atoms.

Questions

1 Aqueous sodium hydroxide solution contains a strongly nucleophilic entity. What is this? Predict the major product formed when ethyl bromide is treated with warm aqueous sodium hydroxide solution and show the electron pair shifts in the reaction.

2 Now give similar answers for the reaction of ethyl bromide with hot aqueous alcoholic potassium cyanide.

3 Place the following acids in order of decreasing acidity:

$Cl \cdot CH_2 \cdot CHCl \cdot COOH$	(A)
$Cl_2 \cdot CH \cdot CH_2 \cdot COOH$	(B)
$CH_3 \cdot CCl_2 \cdot COOH$	(C)
$BrCH_2 \cdot CHBr \cdot COOH$	(D)
$ICH_2 \cdot CHI \cdot COOH$	(E)
$CH_3 \cdot CH_2 \cdot COOH$	(F)

4 State which of the following compounds possess dipole moments: Hydrogen, chlorine, hydrogen chloride, chloroform, benzene, chlorobenzene, *p*-dichlorobenzene, *cis*-dichloroethylene, *trans*-dichloroethylene.

2.3 The iodoform reaction

The iodoform reaction is the well-known diagnostic test for the structural units

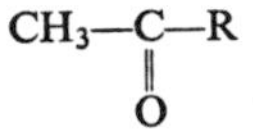

$$CH_3\!-\!\underset{\underset{O}{\|}}{C}\!-\!R \qquad \text{and} \qquad CH_3\!-\!\underset{\underset{OH}{|}}{CH}\!-\!R$$

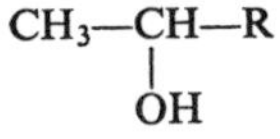

in organic compounds, where R=H or alkyl.

Upon warming a compound containing one of these structural units with a solution of iodine in dilute aqueous alkali, iodoform CHI_3, is produced. This is a pale yellow solid, insoluble in water, with a characteristic odour and melting point.

The simplest ketone to give this reaction is acetone, CH_3COCH_3, and the stages whereby this compound yields iodoform with iodine in dilute aqueous sodium carbonate or sodium hydroxide solution will now be described.

THE STRUCTURE OF ACETONE

The first step in the reaction is governed by the structure of the acetone molecule, notably by the nature of the carbonyl group. There is a covalent double bond between the carbon and oxygen atoms which is made up of four electrons, two being provided by each atom, leaving two lone pairs of electrons on the oxygen atom.

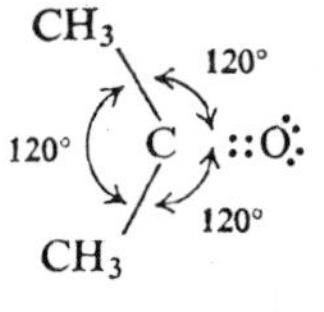

The central carbon atom is in an sp^2 hybridized state, and the bonds round this atom are planar and trigonal with bond angles of $120°$. However, as oxygen is more electronegative than carbon (see table page 9), the electrons forming the carbon–oxygen double bond do not distribute themselves evenly between the two nuclei. There is a greater electron density on oxygen compared with carbon. The true structure of the carbonyl group lies some-where between the written forms shown:

$$\text{>C=\underset{..}{\overset{\curvearrowright}{O}}:} \leftrightarrow \text{>}\overset{\oplus}{C}\text{--}\overset{..}{\underset{..}{O}}\text{:}^{\ominus}$$

Acetone therefore is more truly represented by the structure

$$\begin{array}{c} CH_3 \\ \diagdown\, \delta+ \quad \delta- \\ \qquad C\!\!=\!\!O \\ \diagup \\ CH_3 \end{array}$$

and its dipole moment of 2·8 D is in complete agreement with this formulation.

The carbon–oxygen bond is therefore polarized to a marked degree and has a partial ionic character, resulting in electron deficiency on the central carbon atom.

PROTON ABSTRACTION AND RESONANCE

It is possible for a base $:B^{\ominus}$ under suitable conditions to abstract a hydrogen atom (as a proton) from a carbon atom adjacent to a carbonyl group. Such hydrogen atoms, termed α (next door) hydrogen atoms, are said to be 'active' and they are very slightly acidic.

$$\alpha\text{-carbon atom} \quad \overset{|}{\underset{\underset{B:^{\ominus} \longrightarrow H}{|}}{C}}\!\!-\!\!\overset{|}{\underset{O}{C}}\!\!- \;\rightarrow\; -\overset{|}{\underset{..}{C}}\!\!-\!\!\overset{|}{\underset{O}{C}}\!\!-\!+H:B$$

(I)

The driving force for this sequence comes from two sources. Firstly, the electronegativity-induced pull of electrons by the carbonyl oxygen atom is transmitted to the α-carbon atom; electron attraction away from the α-carbon atom will tend to weaken the α-carbon–hydrogen bond. A concerted electron pair shift clearly depicts this.

$$-\overset{|}{\underset{\underset{B:^{\ominus} \longrightarrow H : \underset{..}{\overset{..}{O}}}{|}}{C}}\!\!-\!\!\overset{|}{C}\!\!- \;\rightarrow\; -\overset{|}{C}\!\!=\!\!\overset{|}{C}\!\!-\!+H:B \quad :\underset{..}{\overset{..}{O}}:^{\ominus}$$

(II)

The anion produced by this proton abstraction can have either of the written structures I or II, and herein lies the source of the second driving force for the abstraction. The extra electron 'made free' by proton removal, which gives the anion its unit negative charge, can be formally accommodated at two sites: on the α carbon atom (I), or on the oxygen atom (II). The true structure of the anion thus lies somewhere between the two written structures or 'canonical forms' which are said to be '*in resonance*':

(I)   (II)

(The arrows ↔ depict resonance.)

The unit negative charge (one electron) does not 'reside' on the carbon atom where it was originally received when deprotonation took place, but it is spread out or 'delocalized' over the whole three-atom system. The anion is said to be '*stabilized by resonance*'. It should be noted that in practice there is only *one* structure for the anion, and chemists equate 'resonance' with electron delocalization rather than electronic oscillation. An analogy to resonance is if we imagine a system of several people and a hot metal ball, the people being the atoms of a resonant system and the hot ball being an extra acquired electron. It is possible to prevent the hot ball from falling to the ground if it is continually quickly passed from person to person in the group. The system is 'stabilized' because the ball can be shared out among a number of pairs of hands; one person could not hold the ball for long without acute discomfort; the ball would be dropped and the system would have broken apart. Just as in electrostatics, the larger the area over which a given change is spread, the lower its potential, so in chemistry, resonance delocalization allows more space for the electrons and thus lowers the energy of the system. Because the extra electron in the anion above is shared out over three atoms, the anion has a greatly enhanced stability over an anion structure which would require localization of the negative charge on one particular atom. The enhanced stability of the potential anion is therefore a positive driving force for the deprotonation of hydrogen atoms α to a carbonyl group. Ethane, for example, shows no tendency to deprotonate under comparable conditions because the potential anion requires localization of the negative charge on one atom. No other canonical forms can be written; the spare unshared electron pair has 'nowhere to go' without increasing the number of valence electrons round one carbon atom beyond the permitted maximum of eight.

Several other examples of resonance stabilization will be encountered later.

In solutions of sodium hydroxide or sodium carbonate, hydroxyl ion is the most predominant strongly basic entity, being generated in the case of sodium carbonate by the hydrolysis of carbonate ions.

$$CO_3^{\ominus\ominus} + H—OH \rightleftharpoons HCO_3^{\ominus} + OH^{\ominus}$$
$$HCO_3^{\ominus} + H—OH \rightleftharpoons H_2CO_3 + OH^{\ominus}$$

The first stage of our iodoform reaction is therefore proton abstraction from the acetone molecule by hydroxyl ion to yield a resonance stabilized anion.

$$H—\ddot{\underset{..}{O}}{:}^{\ominus} \longrightarrow H{\vdots}CH_2—CO—CH_3 \rightarrow H_2O + :\overset{\ominus}{C}H_2—CO—CH_3$$

IODINATION

This anion is however not likely to have a long existence. If it is approached by an iodine molecule in solution, then it will polarize the iodine molecule by tending to repel electrons from the iodine atom which is nearer to the anion's unshared pair of electrons. This stimulates nucleophilic attack by the unshared

$$\underset{\substack{\text{Approaching} \\ \text{iodine molecule} \\ \text{(polarized)}}}{\overset{\delta- \quad \delta+}{I—I}} \longleftarrow \overset{\ominus}{:}CH_2—CO—CH_3 \rightarrow I:^{\ominus} + I—CH_2—CO—CH_3$$

Approaching   (I)   monoiodo ketone

electron pair on to the positive end of the iodine molecule. The stable iodide anion is liberated and monoiodoacetone is the intermediate product. Alternatively, the iodine molecule can be pictured as becoming polarized as it approaches the electron rich carbon–carbon double bond (see page 52) of anion structure II to yield the same products by a concerted electron shift.

$$\begin{array}{c} I^{\delta-} \\ | \\ I^{\delta+} \\ | \\ CH_2{=}C—CH_3 \\ | \\ :\ddot{O}{:}^{\ominus} \end{array} \rightarrow I^{\ominus} + \begin{array}{c} I \\ | \\ CH_2—C—CH_3 \\ \| \\ O \end{array}$$

(II)

Rate of reaction studies on the halogenation of ketones confirm the two stage process described. In alkaline solution, the halogenation of acetone obeys the rate expression

Rate α [acetone] [base]

and is independent of [halogen]. Thus, the reaction between acetone and the base, the deprotonation stage to form the resonance stabilized anion, is the slow rate-determining or 'bottleneck' step which holds up the relatively fast iodination stage:

$$CH_3-\underset{\underset{O}{\|}}{C}-CH_3 \xrightleftharpoons{OH^\ominus,\ slow} \left[\ \overset{\ominus}{C}H_2-\underset{\underset{O}{\|}}{C}-CH_3 \leftrightarrow CH_2\!=\!\underset{\underset{O^\ominus}{|}}{C}-CH_3\ \right] + H_2O$$

$$\downarrow \begin{matrix} I_2 \\ fast \end{matrix}$$

$$I-CH_2-\underset{\underset{O}{\|}}{C}-CH_3 + I^\ominus$$

In the presence of sufficient iodine, further iodination of monoiodoacetone now takes place by a similar mechanism. Extra iodine atoms are introduced for hydrogen preferentially at the carbon atom which has already been substituted. This is because the presence of the iodine atom in monoiodoacetone enhances the acidity of the adjacent hydrogen atoms in the presence of a base, by drawing electrons away from the C—H bond, and thus favouring the vital deprotonation step. Electron withdrawal from this hydrogen atom is occurring in *two* directions.

$$I \longleftarrow \overset{H\diagdown \overset{..}{\underset{..}{O}}{}^\ominus}{\underset{}{\overset{\downarrow}{H}}}{}\ \underset{\underset{O_{\delta-}}{\|}}{\overset{\delta+}{C}}-CH_3 \rightleftharpoons H_2O+I-\overset{\ominus}{C}H-\underset{\underset{O}{\|}}{C}-CH_3$$

Resonance stabilized
carbanion

In addition, this process is favoured over deprotonation from the methyl group because the large volume of the iodine atom will help to stabilize the resulting carbanion. Rapid reaction as before of this carbanion with iodine molecules yields the 1, 1 disubstituted intermediate.

$$I-\overset{\overset{I-I}{\underset{}{}}}{\underset{\underset{O}{\|}}{C}}\overset{\ominus}{H}-\underset{\underset{O}{\|}}{C}-CH_3 \rightarrow I-\overset{\overset{I}{|}}{C}H-\underset{\underset{O}{\|}}{C}-CH_3 + I^\ominus$$

In a similar fashion, this compound reacts with more iodine to yield $CI_3 \cdot CO \cdot CH_3$. It is not possible to halt the iodination reaction before this stage in alkaline solution.

HYDROLYSIS OF $CI_3 \cdot CO \cdot CH_3$

The final stage of the iodoform reaction involves nucleophilic attack by hydroxyl ion on the electron deficient carbon atom of the carbonyl group of tri-iodoacetone. The electron deficiency on this central carbon atom in the presence of a nucleophile is caused by a two-fold inductive pull of electrons by the iodine atoms and the oxygen atom. An —OH group is therefore substituted for —CI_3 which is expelled as an unstable anion:

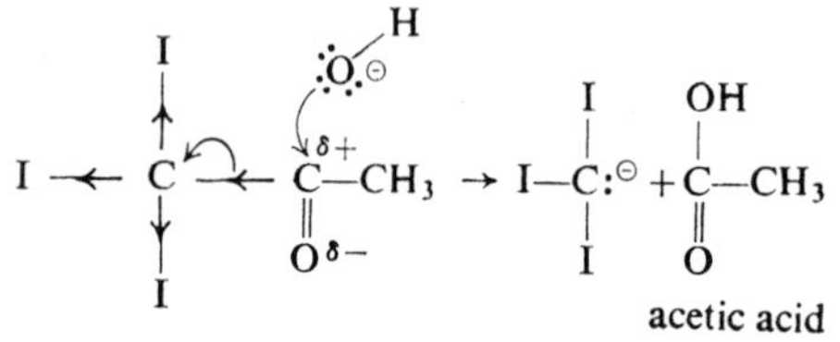

Protonation of the unstable $^{\ominus}CI_3$ ion (which is a very strong base) rapidly takes place from a suitable species in solution, for example the acetic acid formed in the reaction above, to yield iodoform.

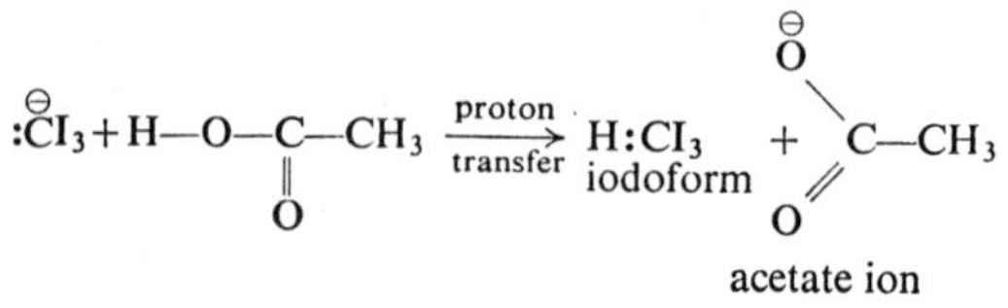

This last stage of the reaction is favoured because the acetate ion is stabilized by resonance, the negative charge being formally accommodated on either oxygen atom.

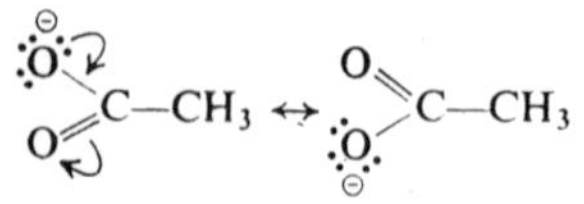

In the sodium carbonate solution therefore sodium acetate will be one of the products, and the overall reaction equation can be summarized:

$$CH_3COCH_3 + 3I_2 + 4NaOH \rightarrow CHI_3 + CH_3COONa + 3NaI + 3H_2O$$

the sodium hydroxide either being added as such in dilute solution, or being generated by salt hydrolysis of sodium carbonate in aqueous solution. It should be emphasized that the reaction does not proceed quantitatively according to the above combined equation.

IODOFORM FROM ALCOHOLS

When iodoform is obtained from alcohols with the functional unit CH_3—$CH(OH)$—R by means of alkaline iodine (hypoiodite) solution, the first stage of the reaction is oxidation of the alcohol to the corresponding carbonyl compound, for example:

$$CH_3\text{---}CH(OH)\text{---}CH_3 \xrightarrow{-2H} CH_3\text{---}CO\text{---}CH_3$$

$$CH_3\text{---}CH_2OH \xrightarrow{-2H} CH_3CHO$$

The mechanism of this oxidation is not clearly understood, but it can be depicted as proceeding via an alkyl hypoiodite, followed by elimination of the elements of hydrogen iodide, for example:

$$CH_3\text{---}CH_2\text{---}OH + i_2 \rightarrow CH_3CH_2OI + (HI)$$

then

$$\underset{\substack{|\\H}}{\overset{CH_3}{H\text{---}C}}\text{---}O\text{---}I \rightarrow H\text{---}\underset{}{\overset{CH_3}{C}}\text{==}O + (HI)$$

Thereafter, the reaction sequence on acetaldehyde (R=H) or the methyl ketone is entirely similar to the reaction described for acetone; namely, the successive deprotonation of three active hydrogen atoms of one methyl group, followed by successive replacement by iodine, and alkaline hydrolysis of the tri-iodo derivative to iodoform and a carboxylate anion.

PRINCIPLES INTRODUCED IN THIS SECTION

1. **Resonance.** *When two or more valid electronic structures can be written for a chemical entity, the actual structure lies somewhere between the written forms. Electron pair shifts within each written structure convert them into one of the alternatives. This delocalization, or 'smearing out' of the bonding electrons within a chemical species is termed resonance. If resonance is possible for an entity, its stability is enhanced.* (See also section 2.5.)

2. **Neighbouring group effects.** *Isolated paraffinic C—H bonds are very strong, making the paraffins a fairly inert series of compounds. They can however be relatively easily ruptured if they possess an aldehyde or ketone group as an adjacent neighbour. The influence of one bond or grouping upon another suitably situated within the same molecule is a common feature in organic chemistry.*

22

Questions

1 What are the three major steps in the production of iodoform from ethyl alcohol with iodine in dilute sodium hydroxide solution? Name and give the formula of the carboxylic acid salt formed in the reaction.

2 Which of the following compounds give a positive iodoform test?

$CH_3.CHO$	A
$CH_3.CH_2 \cdot CHO$	B
$CH_3.CH_2.CH(OH).CH_2.CH_3.$	C
$CH_3.CH_2.CH_2.CH(OH).CH_3.$	D
$(CH_3)_3C.CO.CH_3.$	E

3 How would you obtain chloroform, $CHCl_3$, from acetone?

2.4 Esterification

Reaction of a carboxylic acid, RCOOH, with an alcohol, R'OH, in the presence of an acid catalyst yields, fairly rapidly on heating, an equilibrium mixture with an ester, RCOOR', and water. A common example of this reaction is the laboratory preparation of ethyl acetate.

$$CH_3COOH + CH_3CH_2OH \overset{H^{\oplus}}{\rightleftharpoons} CH_3COOCH_2CH_3 + H_2O$$
ethyl acetate

The most common acid catalysts, present to a few per cent in solution, are dry hydrogen chloride or concentrated sulphuric acid. These accelerate the rate of attainment of equilibrium, though the latter probably shifts the equilibrium to the right by removing the water formed. An excess of one of the reagents is frequently used to obtain a higher percentage conversion of the other reagent to the ester. Without the acid catalyst, the reaction proceeds extremely slowly even at elevated temperatures, and so the acid, or more strictly, protons, play an important role in the reaction.

THE CARBOXYL GROUP

This functional unit is a combination of a carbonyl group and a hydroxyl group on the same carbon atom. The closeness of these two groups results in a considerable interaction, so that carboxylic acids have markedly different properties from ketones and alcohols. This is due to the significant tendency of these acids to ionize:

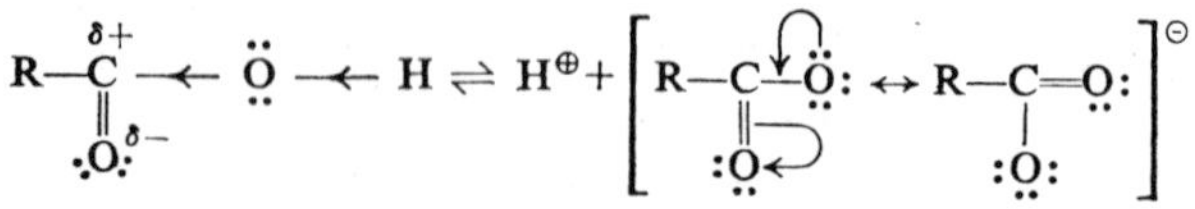

This ionization, a proton release, is favoured by the inductive pull of electrons away from the O—H bond by the electron deficient carbonyl carbon atom, and by the resonance stabilization of the resulting anion. These factors do not operate in alcohols making proton release much more difficult in these compounds.

$$CH_3-CH_2-\overset{..}{\underset{..}{O}}-H \rightleftharpoons [CH_3-CH_2-\overset{..}{\underset{..}{O}}:]^{\ominus}+H^{\oplus}$$

no other significant
contributing structures

Acetic acid is in fact about 10^{11} times as strong an acid as ethanol:

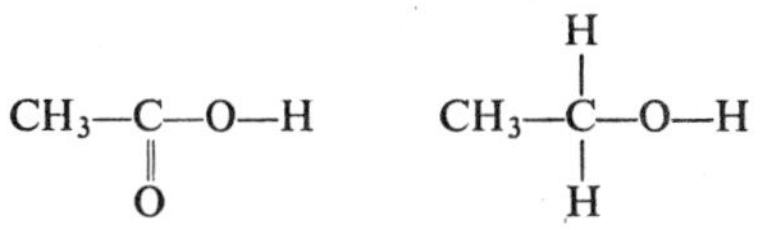

$$\underset{\underset{pK_a\ 4\cdot8}{\overset{\parallel}{O}}}{CH_3-C-O-H} \qquad \underset{\underset{pK_a\simeq16}{\overset{|}{H}}}{\overset{\overset{H}{|}}{CH_3-C-O-H}}$$

In the presence of a base (nucleophile), the primary tendency for a carboxylic acid will therefore be to ionize:

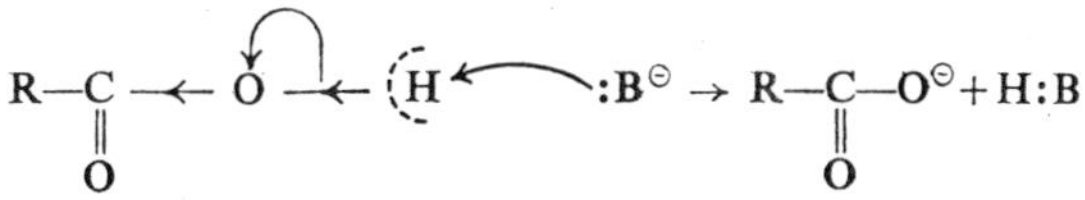

$$\underset{O}{\overset{\parallel}{R-C}}-\overset{..}{O}-H \cdots :B^{\ominus} \rightarrow \underset{O}{\overset{\parallel}{R-C}}-O^{\ominus}+H:B$$

The alternative reaction in which the base attacks the slightly electron deficient carbonyl carbon atom of the unionized acid is much less likely.

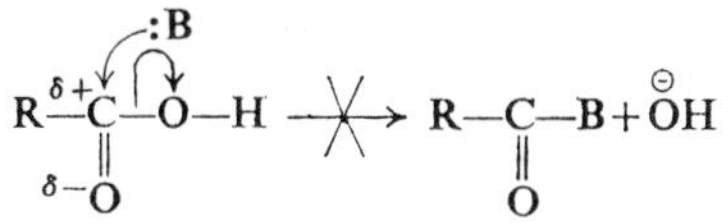

$$\underset{\delta-O}{\overset{:B}{R\overset{\delta+}{-}C-O-H}} \xrightarrow{\quad\times\quad} \underset{O}{\overset{\parallel}{R-C}}-B+\overset{\ominus}{O}H$$

Moreover, once the resonance stabilized carboxylate anion has formed, it will have a natural repulsion for nucleophilic entities, and it will therefore be even less likely to be attacked by them. Thus, the carbonyl carbon atom of a carboxylic acid is not in practice anything like so electron deficient as that of a ketone, and herein lies the main reason for their difference in properties. (See also section 3·2.)

PROTONATION BY THE ACID CATALYST

When the acid catalyst is added to a mixture of the alcohol and the carboxylic acid, a series of proton equilibria are established, the two most important for the esterification reaction being those with the reactants:

$$CH_3-\underset{\underset{\overset{..}{O}}{\overset{\|}{}}}{C}-\overset{..}{\underset{..}{O}}-H + H^{\oplus} \rightleftharpoons CH_3-\underset{\underset{\overset{..}{O}}{\overset{\|}{}}}{C}-\overset{\overset{\displaystyle H}{\diagup}}{\underset{\diagdown}{\oplus O:}}_{H} \rightleftharpoons CH_3-\underset{\underset{H}{\overset{|}{:\overset{\oplus}{O}}}}{\overset{\|}{C}}-\overset{..}{O}-H$$

$$\text{(I)} \qquad\qquad\qquad\qquad\qquad \text{(II)}$$

$$CH_3-CH_2-\overset{..}{\underset{..}{O}}-H + H^{\oplus} \rightleftharpoons CH_3-CH_2-\overset{\overset{\displaystyle H}{\diagup}}{\underset{\diagdown}{\oplus O:}}_{H}$$

These oxonium cations are formed by lone pair donation by oxygen. In the case of protonation of the carboxylic acid, two structures can result, these also being in equilibrium as shown (I and II). Both (I) and (II) have a formal positive charge on oxygen which will attract electrons from the carbonyl carbon atom making this atom considerably more electron deficient than it was in the unprotonated form. Thus, protonation of the carboxyl group has increased the ability of the carbonyl carbon atom to accept electrons from a suitable nucleophile.

ISOTOPIC TRACER STUDIES

The use of isotopic labelling of alcohols has shed considerable light on the mechanism of the reaction. In our example, ethyl alcohol enriched with O^{18}, when reacted with acetic acid, was found to generate water in the reaction with no O^{18} enrichment. The O^{18} was found in the ester. Hence in the overall reaction the acyl-oxygen bond of the acid must break and the alkyl–oxygen bond of the alcohol must be preserved, that is:

$$CH_3-\underset{\underset{O}{\overset{\|}{}}}{C}\Big|-O-H + CH_3.CH_2-O^{18}\Big|-H \rightarrow CH_3-\underset{\underset{O}{\overset{\|}{}}}{C}-O^{18}-CH_2CH_3 \atop +H_2O$$

A joining of the oxygen atom of the alcohol to the carbonyl carbon atom of the acid must therefore take place at some time during the reaction.

THE REACTION PATH

Of the oxonium cations available for reaction, protonated acetic acid structure (I) is the most suitable. Protonated ethanol, with its weak C—O bond, cannot be an important entity in the reaction, since the alcohol carbon–oxygen bond must be preserved. Protonated acetic acid structure (II) has two resonance forms

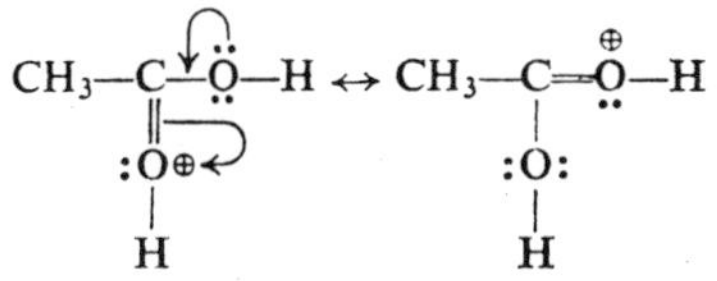

$$CH_3-C-\ddot{O}-H \leftrightarrow CH_3-C=\overset{\oplus}{O}-H$$

$$:\overset{\oplus}{O} \qquad\qquad :O:$$

$$H \qquad\qquad\quad H$$

(II)

which will give it enhanced stability over structure (I). In addition, structure (I) has, in a neutral water molecule, a good potential leaving group in acid solution if its carbonyl carbon atom were attacked by a nucleophile.

In the reaction solution, only neutral ethanol molecules will serve as nucleophilic attacking species on structure (I) to give rise to the ester with the correct new C—O bond. The reaction path is therefore one of nucleophilic attack by ethanol on the electron deficient carbonyl carbon atom of structure (I):

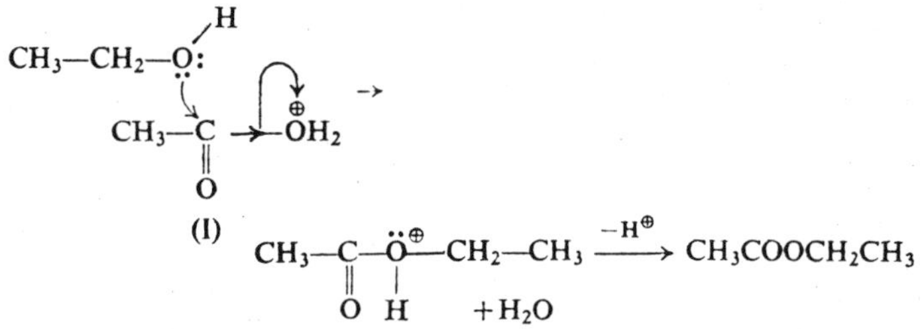

$$CH_3-CH_2-\ddot{O}:^H$$

$$CH_3-C \rightarrow \overset{\oplus}{O}H_2 \quad\rightarrow$$

$$O$$

(I)

$$CH_3-C-\overset{\oplus}{\ddot{O}}-CH_2-CH_3 \xrightarrow{-H^{\oplus}} CH_3COOCH_2CH_3$$

$$O \quad H \qquad +H_2O$$

The last step of proton loss will occur on to any other proton acceptors in solution. All the stages in this reaction are reversible equilibria so that the series of equations can represent the acid-catalyzed hydrolysis of ethyl acetate as well as the acid catalyzed esterification of acetic acid.

REACTION OF ALCOHOLS WITH ACID CHLORIDES

Esters can be rapidly produced by the action of alcohols on acid chlorides. No catalyst is required because the carbonyl carbon atom of the acid chloride, as opposed to that of an acid, is distinctly electron deficient due to the inductive pull of electrons by both oxygen and chlorine.

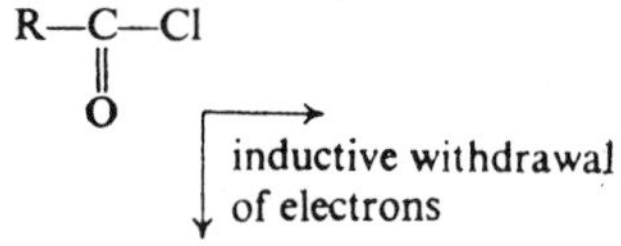

$$R-C-Cl$$

$$O$$

inductive withdrawal of electrons

Nucleophilic attack by an alcohol molecule on to this electron deficient carbon atom can therefore easily take place.

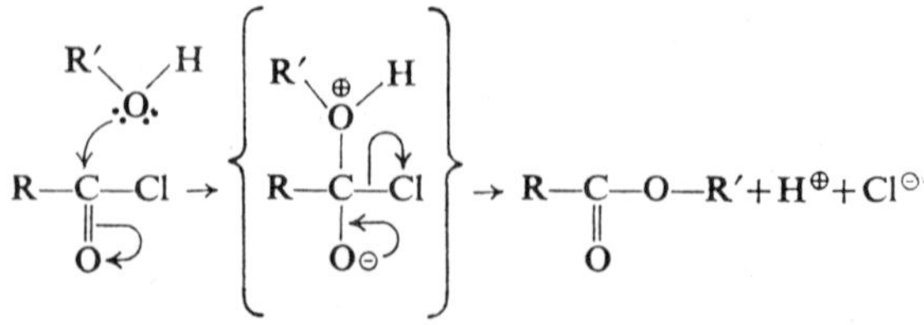

A transitory addition intermediate is shown here in brackets; this expels $Cl^{\ominus}$ and then deprotonates to form the ester. (It could alternatively expel $R'OH$, but this course would be of no consequence since it would yield the starting reactants.) This type of intermediate is often shown for clarity in many reactions which involve nucleophilic displacement at a carbonyl carbon atom*. (See also section 3·3.)

Questions

1 Esters are hydrolyzed by warm aqueous sodium hydroxide to the alcohol and the sodium salt of the carboxylic acid from which they were compounded†. Isotopic labelling indicates that the acyl–oxygen bond of the ester is ruptured during reaction, for example:

$$\underbrace{CH_3CO}_{\substack{\text{the}\\\text{acyl}\\\text{group}}}-O^{18}-CH_2CH_3 + Na^{\oplus}OH^{\ominus} \rightarrow CH_3COO^{\ominus}Na^{\oplus} + HO^{18}CH_2CH_3$$

Formulate the reaction path of the alkaline hydrolysis of ethyl acetate giving the intermediate addition entity. (Hint: the carbonyl carbon atom of the ester is slightly electron deficient.)

2 Acetic anhydride reacts with ethanol rapidly at about 60°C to yield ethyl acetate and acetic acid. Formulate the reaction sequence in terms of intermediates and electron pair shifts.

* If the incoming nucleophile binds itself completely to the carbonyl carbon atom before the group to be expelled begins to break off, the reaction is energetically favoured. By such a route, the system receives much of its 'energy payment' from the formation of the new bond to carbon before having to pay its 'energy debt' for the breaking of the old bond to carbon, and the energy barrier (activation energy) of the reaction is thereby reduced.

† An excellent colour film on the mechanism of ester hydrolysis (for methyl benzoate) can be hired from Sound Services Film Library, Wilton Crescent, London, S.W.19. Reference number 41/66.

2.5 Aromatic substitution

In general terms, aromatic substitution involves the replacement of an atom or group, usually hydrogen, attached to an aromatic ring system by another atom or group of atoms. Most simply, for benzene, C_6H_6, the reaction would be of the type

$$C_6H_6 + XY \rightarrow C_6H_5X + HY$$

This reaction appears, stoichiometrically, to be similar in type to a normal aliphatic substitution reaction. The mechanism and conditions by which it occurs are however generally markedly different.

The most simple aromatic compound, benzene, has its basic carbon skeleton embodied in the great majority of aromatic substances, and in order to understand the mechanics of aromatic substitution we must first briefly consider its structure.

THE STRUCTURE OF BENZENE

Benzene has six equivalent carbon atoms joined together as a planar ring. The six hydrogen atoms are also equivalent, one being attached to each carbon atom. Each carbon atom is therefore bound to three other atoms by normal covalent bonds which have their centre of electron density along the line joining the nucleii of the atoms. These bonds, termed sigma (σ) bonds, are the result of sp^2 hybridization, or orbital mixing, of the valence electrons of carbon, which yields trigonal planar bonds.

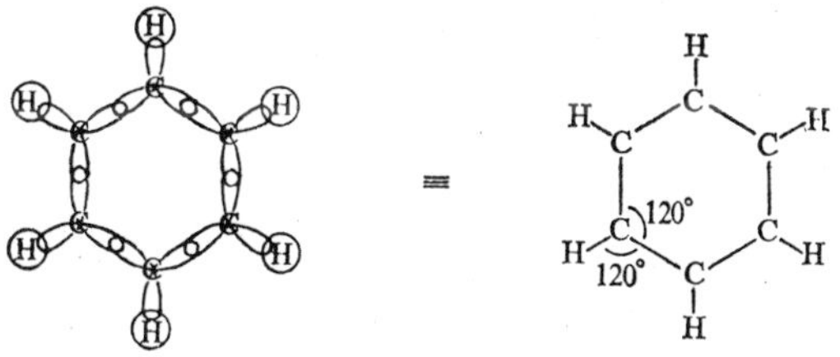

Planar benzene structure showing σ bonds only.

The σ bonds shown however only involve three of the four valence electrons of each carbon atom. There are therefore a total of six 'spare' electrons available for further bonding, one being associated with each carbon atom.

These electrons occupy the $2p_z$ orbital of carbon which extends above and below the plane of the ring of carbon atoms:

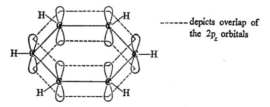

------ depicts overlap of the $2p_z$ orbitals

The $2p_z$ electrons achieve bonding by overlapping their orbitals to form annular charge clouds above and below the plane of the ring. These bonding electrons which have their charge clouds outside the line joining the atomic nucleii are called pi (π) electrons.

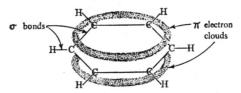

σ bonds — — — — π electron clouds

This is the most accurate structure for benzene, but it is difficult to depict by means of conventional bond lines so that each carbon atom has four bonds. The Kekulé structures are perhaps the most satisfactory, and will be used in the remainder of this book. These represent benzene with alternate single and double bonds in the ring in two ways.

(I)   (H)

The established shorthand notation for these being

(I)   (II)

The actual structure lies between these two formal possibilities, all the carbon–carbon bonds being identical and of length 1·39 Å. The intermediate nature of these bonds is readily seen, for the carbon–carbon single bond length (for example, in ethane) is 1·54 Å and the carbon–carbon double bond length (for example, in ethylene) is 1·34 Å. The carbon–carbon bonds in benzene are therefore something between 'single' and 'double' in type.

THE REACTIVITY OF BENZENE

Although from the molecular formula and structure of benzene we see that the molecule is unsaturated by some three units, it displays a remarkable resistance to addition reactions characteristic of olefins (see section 3·1). It does not, for example, decolourize bromine water or potassium permanganate solution at room temperature. Most of its reactions are substitutions; the benzene ring resists saturation.

The stability of the aromatic ring is explained by the concept of '*resonance energy*'. The 'resonance energy' is the *difference* in energy content of benzene compared with that of a formal written structure. Thermochemical measurements give us a tangible insight into this concept of resonance energy.

Calorimetric measurement of the heats of hydrogenation of olefinic compounds gives the following results (in kcal/mole):

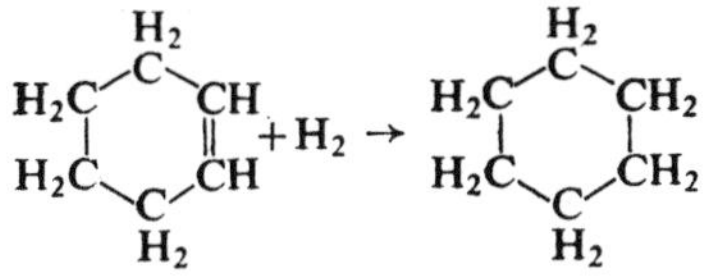

cyclohexene
(one $>C=C<$)

cyclohexane
$\Delta H = -28\cdot5$

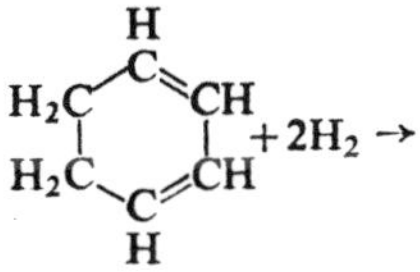

1,3 cyclohexadiene
(two $>C=C<$)

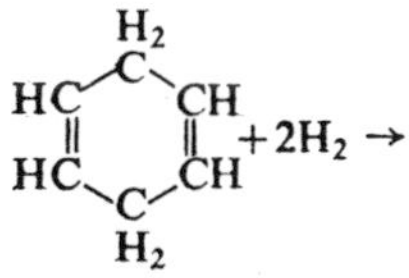

$+2H_2 \rightarrow$ cyclohexane
$\Delta H = -55\cdot5$

1,4-cyclohexadiene
(two $>C=C<$)

$+2H_2 \rightarrow$ cyclohexane
$\Delta H = -60\cdot5$

These results show that saturation of one olefinic bond with hydrogen liberates on average approximately 29 kcal/mole. On this basis, saturation of a Kekulé structure for benzene with 3 moles of hydrogen to yield cyclohexane, should result in the liberation of approximately $3 \times 29 = 87$ kcal/mole. Measurement of the actual heat of hydrogenation of benzene gives the result, $\Delta H = -50$ kcal/mole. (Benzene will add hydrogen to yield cyclohexane, but more forcing conditions are required than for the hydrogenation of olefins.) Thus, benzene, upon hydrogenation, gives out some 37 kcal/mole *less* than it would do if it possessed a formal Kekulé structure. This is because it is some 37 kcal/mole *more stable* than the Kekulé forms, and this is the value of the 'resonance energy'. The Kekulé structures require localization of the $2p_z$ electrons as specific π bonds alternately between particular carbon atoms. The actual benzene structure has these electrons *de*localized — smeared out — over the whole ring of carbon atoms. In accordance with the earlier discussion of resonance phenomena (section 2·3), we see that the greater the volume over which electrons are accommodated, the higher the stability and the lower the energy of the system.

The second major factor governing the reactivity of benzene arises from the presence of the annular π electron clouds in its structure. These act as a repelling shield to any potential nucleophilic attacking entity, and conversely, as a centre of attraction for electrophilic entities. Thus electrophilic aromatic substitution is by far the more common process, drastic forcing conditions often being required to effect substitution by a nucleophile in the majority of aromatic compounds.

Questions

1 The structure of graphite consists of sheets of carbon atoms joined as edge to edge hexagons. Successive sheets are held together by Van-de-Waals forces and are separated by about 3·5 Å. Each carbon–carbon bond length within a sheet of hexagons is 1·42 Å long. Comment on the type of chemical bonding which carbon atoms display in graphite and suggest the origin of graphite's conducting properties.

2 Would you expect the cyclohexadienes shown on page 29 to decolourize bromine water at room temperature? Give your reasons.

3 Aqueous sodium hydroxide at 60°C will hydrolyze methyl chloride to methanol:

$$CH_3Cl + Na^{\oplus}OH^{\ominus} \rightarrow CH_3OH + Na^{\oplus}Cl^{\ominus}$$

Why are conditions of about 300°C and 200 atmospheres pressure required to convert chlorobenzene, C_6H_5Cl, to phenol, C_6H_5OH, using the same reagent?

A. *Friedel-Crafts reaction*

In the presence of a suitable catalyst, an alkyl halide will react in the cold with benzene to yield a substituted derivative. Methyl chloride gives toluene as the major product.

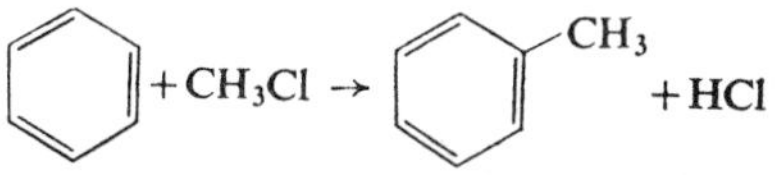

The catalyst however plays a vital role in the reaction. It is responsible for generating a strongly electrophilic attacking entity by taking part in an intermediate reaction with the reagent. Suitable catalysts are all Lewis acids, and anhydrous aluminium chloride is the one most commonly used. In the example, it reacts with the methyl chloride to yield the short-lived, highly electrophilic, methyl carbonium ion, $CH_3^{\oplus}$:

$$CH_3Cl + AlCl_3 \rightleftharpoons CH_3^{\oplus} + AlCl_4^{\ominus}$$

This entity attracts a pair of electrons from the π cloud of the aromatic ring to form an addition intermediate with a new localized σ bond at the carbon atom where the methyl group becomes attached. This carbon atom acquires a tetrahedral sp^3 hybridized configuration.

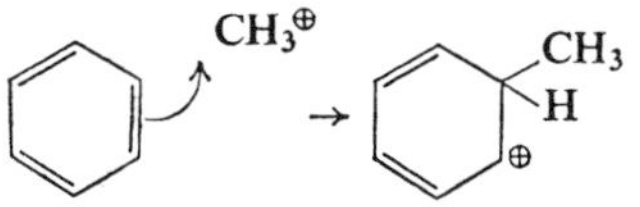

addition intermediate

The addition intermediate is stabilized to some extent because it can share out its unit positive charge among the other five carbon atoms by resonance. Three written structures, which are the result of formal electron pair shifts, are shown:

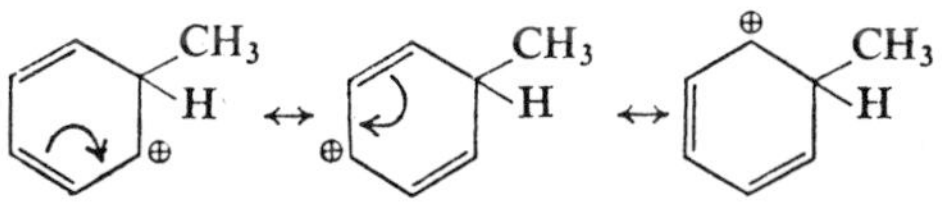

A more satisfactory representation of the addition intermediate which summarizes these structures is

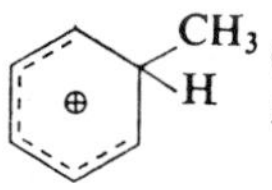

(the dotted line represents four π electrons)

The next stage of the reaction is one of deprotonation of the addition intermediate. In this way, the completely delocalized stability of the six characteristic π electrons of the aromatic ring is restored. The carbon atom where the methyl group is now attached, reverts to its initial state of planar sp^2 hybridization.

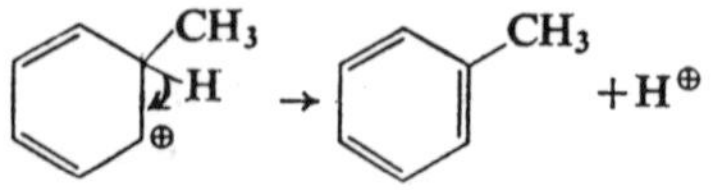

It is possible that the positively charged addition intermediate could combine with an available nucleophilic anion to form a neutral addition compound. This is however far less likely, since this process would involve almost complete loss of the resonance stabilization energy of the aromatic ring, a substituted cyclohexadiene being produced. *Chemical reactions generally favour the most stable product.*

The total reaction can be depicted by a series of concerted electron pair shifts

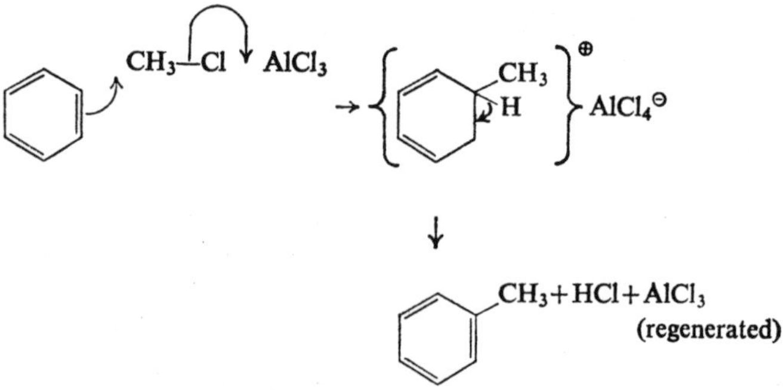

Closely controlled conditions are required to prevent significant further substitution of methyl groups at other positions in the ring.

This mechanism of (a) interaction of the reagent with a suitable catalyst (b) electrophilic attack on the aromatic ring, and (c) deprotonation, is common to other substitution reactions of benzene.

Questions

1 Acid chlorides participate in the Friedel-Crafts reaction. Give the formula of the product produced from the reaction of benzene with acetyl chloride, CH_3COCl, in the presence of aluminium chloride. What is the attacking electrophile?

2 Chlorobenzene is produced from benzene by means of dry chlorine in the presence of finely divided iron as catalyst ('halogen carrier'). What is likely to be the true catalyst and the attacking electrophile?

B. *The nitration of benzene*

Treatment of benzene with a mixture of concentrated nitric and sulphuric acids at a temperature between 30 and 50°C yields nitrobenzene, $C_6H_5NO_2$, as the major product. Reaction of benzene with nitric acid alone fails to produce any significant reaction, and so the concentrated sulphuric acid is a vital catalyst for the introduction of the —NO_2 group into the aromatic ring, it too having virtually no effect on its own on benzene under the above conditions.

Solutions of the two concentrated acids show a four-fold molecular freezing point depression which indicates the presence of four ions. These can be accounted for by the following equilibria:

$$H_2SO_4 + HO\!-\!NO_2 \rightleftharpoons HSO_4^{\ominus} + \left\{ H_2\overset{\oplus}{O}\!-\!NO_2 \right\}$$

$$HSO_4^{\ominus} + H_3O^{\oplus} \xrightleftharpoons[\text{}]{H_2SO_4} \{H_2O\} + NO_2^{\oplus}$$

or overall,

$$2H_2SO_4 + HNO_3 \rightleftharpoons 2HSO_4^{\ominus} + H_3O^{\oplus} + NO_2^{\oplus}$$

The $NO_2^{\oplus}$ (nitronium) ion formed in these equilibria is a powerful electrophile. It can therefore attack the aromatic ring to yield nitrobenzene by an entirely analogous mechanism to the Friedel-Crafts process:

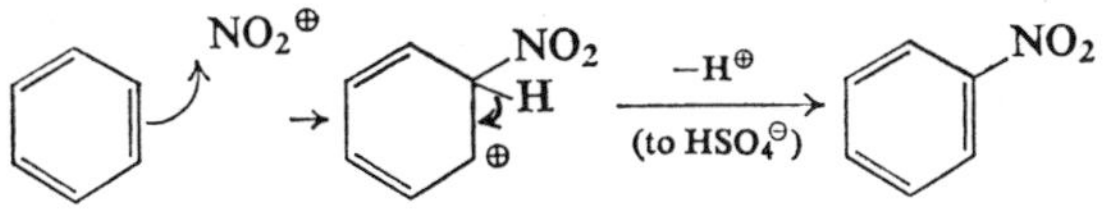

The overall reaction is therefore

$$C_6H_6 + HNO_3 \xrightarrow{H_2SO_4} C_6H_5NO_2 + H_2O$$

If the reaction temperature is about 100°C significant further substitution into the aromatic ring takes place, and metadinitrobenzene can be readily obtained in good yield.

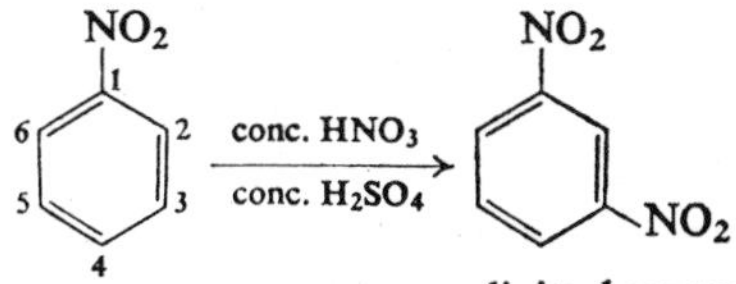

m-dinitrobenzene

The rate at which the second —NO_2 group is introduced into the ring is however found to be slower than the rate at which the first substituent was introduced. Another important aspect of this second substitution is that the second —NO_2 group enters almost entirely (over 90 per cent) into the *meta* position. There are three possible sites at which it could enter, either position 2 (ortho), 3 (meta), or 4 (para), positions 5 and 6 being equivalent to positions 3 and 2 respectively. Why then does the second substituent enter the ring almost exclusively at the meta position? To understand this, we must consider the electron distribution in the nitrobenzene molecule.

THE STRUCTURE OF NITROBENZENE

Nitrogen is a member of the first short period of elements in the periodic table. It cannot therefore expand its shell of valency electrons beyond 8, and this results in a maximum covalency of 4. Organic nitro compounds R—NO_2 must therefore possess a dative covalent link from nitrogen to oxygen

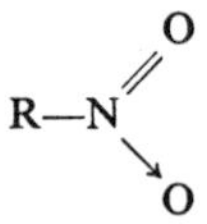

$$R-N\begin{matrix} \nearrow O \\ \searrow O \end{matrix} \quad (\rightarrow \text{ signifies electron pair donation by nitrogen})$$

so that in these compounds the nitrogen atom is formally 4-covalent and there is octet preservation round both oxygen and nitrogen atoms. Both nitrogen–oxygen bonds are however found to be of the same length, and the nitro group is better represented as a resonance hybrid of two canonical forms

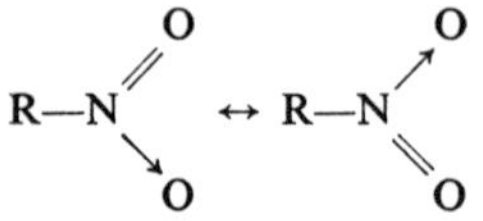

$$R-N\begin{matrix} \nearrow O \\ \searrow O \end{matrix} \leftrightarrow R-N\begin{matrix} \nearrow O \\ \searrow O \end{matrix}$$

the true structure lying between these extreme written forms. An alternative mode of representation which brings out the dipolar character of dative linkages is

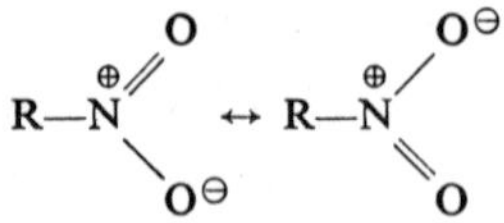

$$R-\overset{\oplus}{N}\begin{matrix} \nearrow O \\ \searrow O^{\ominus} \end{matrix} \leftrightarrow R-\overset{\oplus}{N}\begin{matrix} \nearrow O^{\ominus} \\ \searrow O \end{matrix}$$

The nitrogen atom therefore has a large degree of positive character. The dipole moment of nitromethane (R = CH_3) is 3·1 D, and this confirms the charge separation shown in the above structures.

Nitrobenzene has a dipole moment of 3·95 D indicating a more effective charge separation in this molecule compared with nitromethane. The aromatic ring of nitrobenzene is a potential electron source, and the electron deficiency of the nitrogen atom can be theoretically satisfied by electron pair withdrawal from the ring. As nitrogen cannot become pentavalent, electron withdrawal must formally take place on to oxygen leaving the ring positive. The following canonical forms (ignoring the resonance of the nitro group) arise by the electron pair shifts shown:

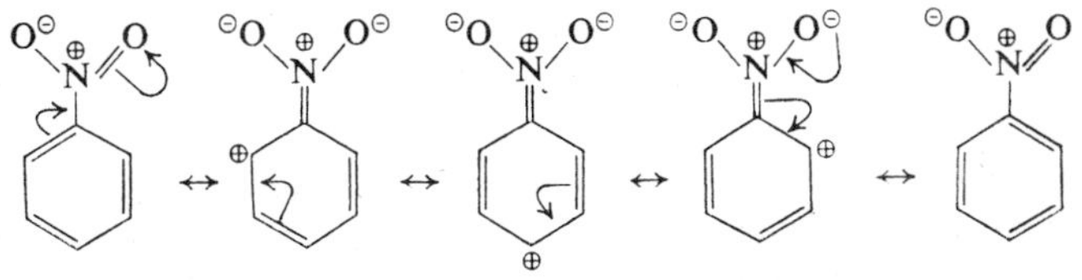

| (I) | (II) | (III) | (IV) | (V) |

Structures (II), (III) and (IV) leave the ring formally positive at the ortho and para positions. It is not possible to draw electron pair shifts on a Kekulé structure which will leave the meta position so charged. The summary structure for nitrobenzene is therefore

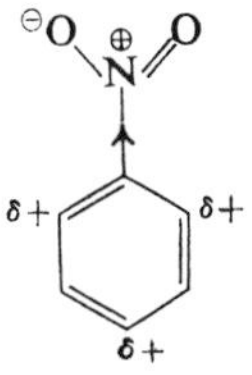

electron withdrawal from the ring by the nitro group leaving the ring slightly electron deficient preferentially at the ortho and para positions.

THE RATE AND POSITION OF ATTACK OF THE SECOND NITRO GROUP

As we have seen, nitration involves attack by the electrophilic $NO_2^{\oplus}$ ion. The polarization shown in the summary structure for nitrobenzene indicates that, compared with benzene, the aromatic ring will be *less* susceptible to attack by electrophiles, such as $NO_2^{\oplus}$, because the ring has become slightly electron deficient due to the electron withdrawing tendency of the nitro substituent. The rate of nitration of nitrobenzene will therefore be slower than for benzene.

The summary structure for nitrobenzene shows that the meta positions are relatively less positive (that is, relatively more negative) than the ortho and para positions. Thus although the electrophilic attacking $NO_2^{\oplus}$ ion is nowhere aided in its attack by the presence of the initial nitro group in the ring, it finds least difficulty in securing electron supply at the meta positions. Hence the product of dinitration of benzene is *m*-dinitrobenzene.

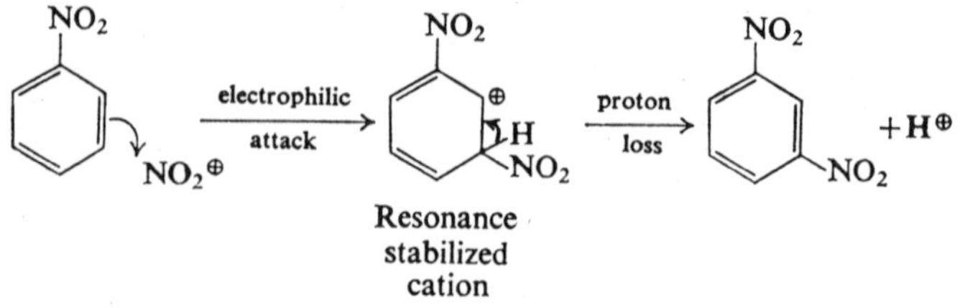

Resonance
stabilized
cation

The introduction of a third nitro group into the benzene ring is hard to accomplish because the two nitro groups of *m*-dinitrobenzene will, by their electron drawing ability, make the ring even less susceptible to attack by an electrophile. The two groups will however reinforce each other making the surviving meta position (which is the same for both groups) relatively least positive. Thus, the predominant further nitration reaction will be

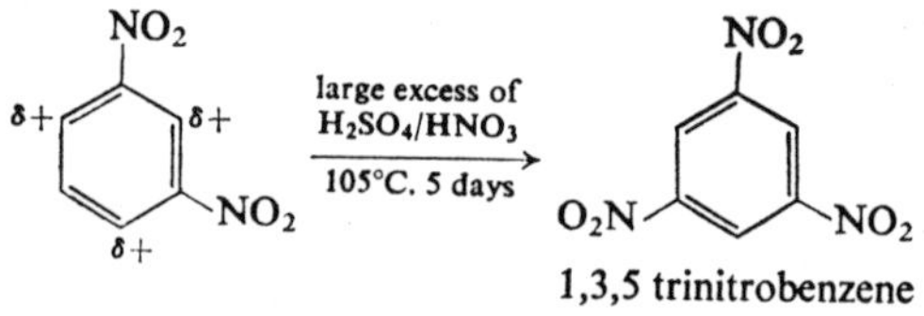

1,3,5 trinitrobenzene

This slow reaction is rarely performed on a preparative scale.

Question

What would you expect to be the product of mononitration of the quaternary amine C_6H_5—$N^{\oplus}$ $(CH_3)_3$ $I^{\ominus}$? Say what you can about the rate of nitration, giving your reasons.

C. The bromination of phenol

When bromine water is added to phenol a pale yellow precipitate of 2, 4, 6-tribromophenol forms almost immediately and the brown colour is discharged.

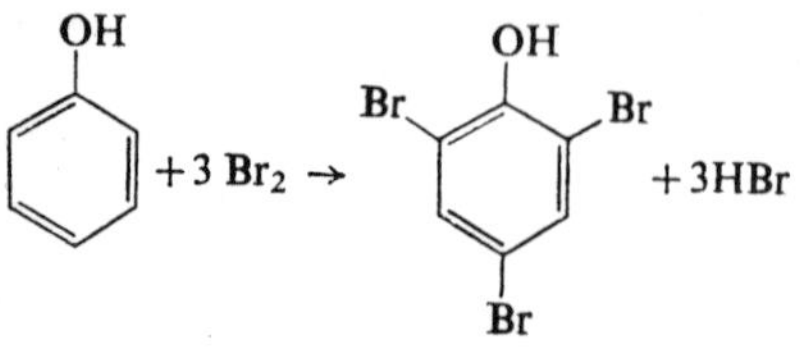

In order to discover why this aromatic substitution is so rapid, and why substitution occurs preferentially at the ortho and para positions, we must consider the electronic structure of phenol more closely.

THE STRUCTURE OF PHENOL

Phenol ($pK_a \simeq 10$) is a much stronger acid than methanol ($pK_a \simeq 16$). The O—H bond of phenol is therefore weaker than that of methanol, and phenol has a significant tendency to ionize to yield a proton and the phenate anion:

$$C_6H_5OH \rightleftharpoons C_6H_5O^\ominus + H^\oplus$$

Once formed, the phenate anion is considerably resonance stabilized by lone pair interaction of the negative oxygen atom with the electron cloud of the aromatic ring, and this favours the ionization.

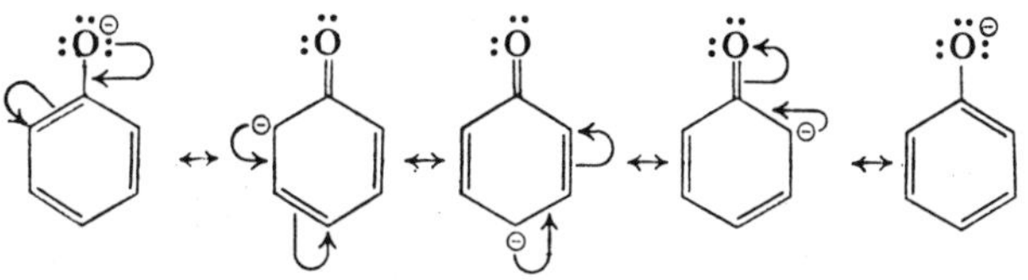

Resonance stabilization also occurs in the undissociated phenol molecule, again by lone pair interaction:

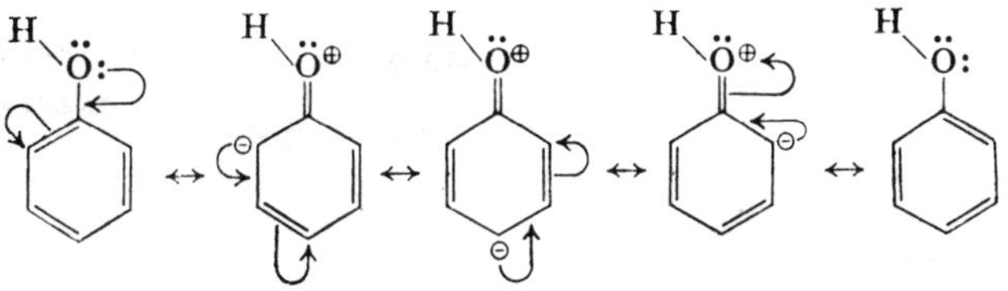

but this is not so effective as in the phenate ion since it involves a separation of unlike charges within the molecule.

Thus, in phenol we have two electronic effects:

(1) an electronegativity induced pull of electrons from the ring towards oxygen. This tends to increase the electron

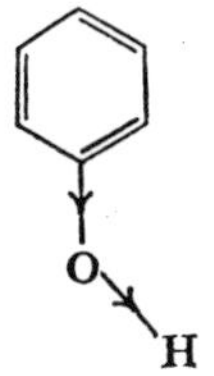

density round oxygen and to increase the strength of the O—H bond, and therefore inhibits proton loss.

and

(2) a resonance induced feed back of electrons from the oxygen lone pairs into the ring. This tends to decrease the electron density round oxygen and to decrease the strength of the O—H bond, and therefore encourages proton loss.

The second effect is clearly the stronger, and overall, the —OH group of phenol is an electron donor to the aromatic ring. Examination of the resonance forms of phenol, or its anion shown above, indicates that electron donation results in formal charges on the ortho and para positions. The summary structure for phenol, which shows the net electron donation effect of the —OH group, is

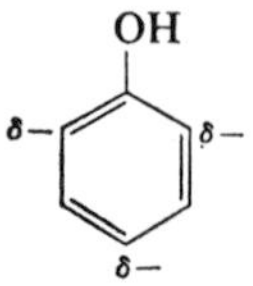

and phenol will therefore be more prone to electrophilic attack than benzene, notably at the ortho and para positions. This accounts for the rapid bromination to yield 2, 4, 6-tribromophenol.

THE REACTION PATH

A bromine molecule approaching a site of high electron density will become polarized by electron repulsion along the Br—Br bond away from the electron rich site. For example, at one of the ortho positions

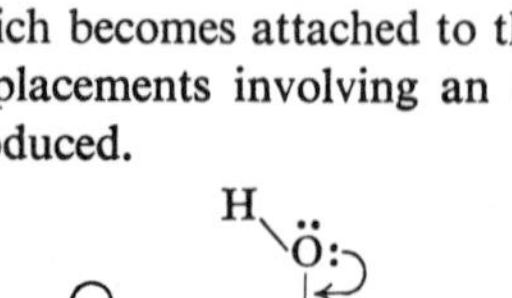

The positive end of the bromine molecule therefore becomes the electrophile which becomes attached to the ring by a series of concerted electron pair displacements involving an oxygen lone pair, hydrogen bromide being produced.

This process is of the same type which we have encountered previously in this section, involving electrophilic attack by the reagent (or, of course, nucleophilic attack by the aromatic centre) to form a resonance stabilized cation, followed by rapid deprotonation. In this reaction we can think of $Br^{\oplus}$ as the attacking electrophile, but it is unlikely that this ion has anything but a very momentary existence during the reaction. Hence the concerted shifts shown are probably the most true representation. Similar formulae show electrophilic demand for electrons by bromine and nucleophilic response by phenol at the para position.

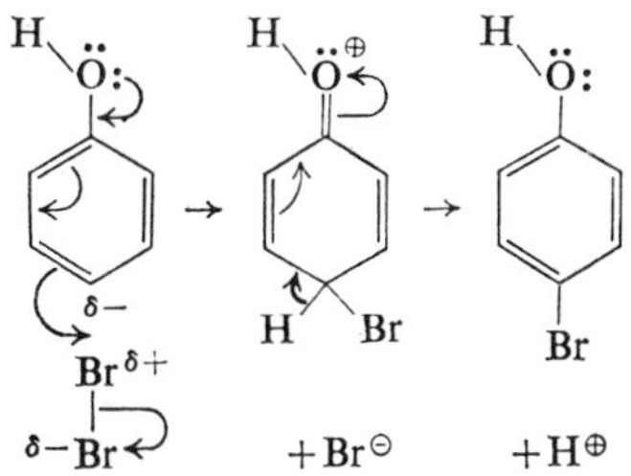

In practice, these reactions take place so quickly, that it is difficult to stop reaction before all three bromine atoms have been introduced into the ring. It should be noted that, in contrast to benzene (see page 32, question 2), no catalyst is required to produce a strongly electrophilic entity for the halogenation of phenol. The inherent electron donation to the aromatic ring by the —OH group, which is augmented as a potential electrophile approaches the ring, is sufficient to cause reaction with fairly weak electrophiles, such as $Br^{\delta+}$—$Br^{\delta-}$, because the ortho and para aromatic centres become that much more electron rich than in benzene.

Questions

1 What can you say about the rate and product of nitration of phenol with excess conc. HNO_3/H_2SO_4? What will be the products of mononitration of phenol using dil. HNO_3?

2 Aniline, $C_6H_5NH_2$, possesses a lone pair of electrons on the nitrogen atom. Giving your reasons, predict the product formed by treating aniline with excess bromine water at room temperature.

3 Amines, which are substituted ammonias, are bases. Predict the relative strengths as bases of aniline and methylamine, CH_3NH_2.

SUMMARY

Substitution into the aromatic ring is most commonly effected by electrophilic entities. These attack the aromatic ring at places of highest electron density to form a resonance stabilized addition intermediate, which subsequently deprotonates to give the substitution product.

In benzene itself all the ring positions are equivalent, but in substituted benzenes, both the position and the rate of reaction are largely controlled by the substituents already present. Substituents, such as —OH and —NH₂, which release electrons to the ring, increase the rate of reaction relative to benzene and direct the incoming electrophile preferentially to the ortho and/or para positions which are said to be 'activated'. Substituents, such as —NO₂ and —N(CH₃)₃, which withdraw electrons from the ring, decrease the rate of reaction relative to benzene and direct the incoming electrophile preferentially to the meta position(s) which are the least 'deactivated'.

The table summarizes the directing effect of common substituents.

Activating groups (Direct mainly ortho/para)	Deactivating groups (Direct mainly meta)
—OH	—NO₂
—NH₂	—NMe₃⁺
—OCH₃	—COOH
—CH₃	—CHO
—Cl*	—SO₂OH
—NHCOCH₃	—CN

*Chlorobenzene directs ortho-para, so that on mononitration a mixture of ortho and para nitrochlorobenzenes results:

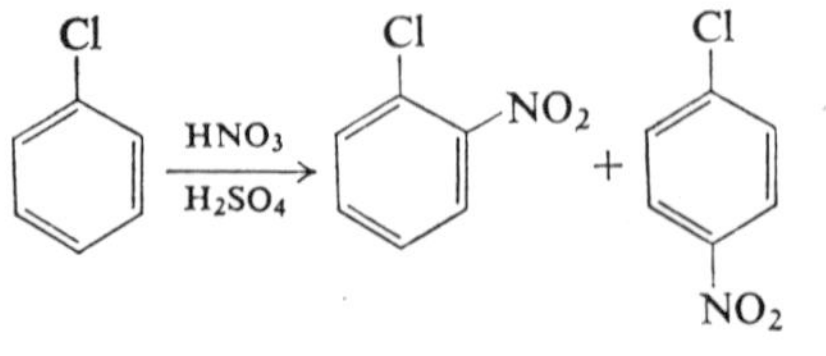

The rate of this reaction is however *slower* than that for benzene under comparable conditions; that is, —Cl is *de*activating. The strongly electronegative chlorine atom exerts a substantial inductive pull of electrons away from the ring (dipole moment 1·55 D), which makes the aromatic ring less susceptible than benzene to attack by the NO₂⁺ ion. However, under the demand of the electrophile as it comes near to the chlorobenzene molecule, lone pair donation by chlorine to the ring can take place by the resonance effect:

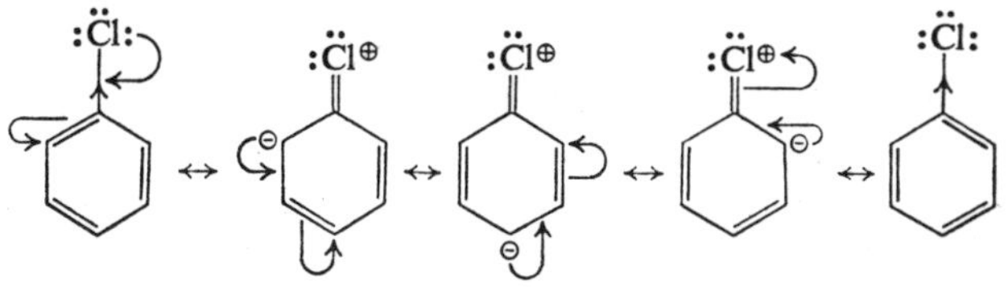

Ortho-para substitution therefore takes place but at a slower rate than for benzene, because under the demand of the reagent, the significant inductive pull from the ring is overcome by the resonance induced feed back into the ring.

2.6 The reaction of amines with nitrous acid

One of the most useful and varied reactions in organic chemistry is that of amines with nitrous acid. The products of the reaction vary considerably, depending upon the structure of the amine and the reaction conditions.

Nitrous acid, which has the structure

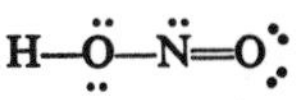

$$H-\ddot{O}-\ddot{N}=\overset{..}{\underset{..}{O}}{\cdot}$$

is an unstable weak acid. It is normally prepared in solution by the addition of a strong acid, such as hydrochloric or sulphuric acid, to a solution of sodium nitrite at 5°C. On ionization, it yields a proton and the resonance stabilized nitrite ion:

$$HNO_2 \rightleftharpoons H^\oplus + NO_2{}^\ominus$$

A. *Primary amines*

The reaction of primary amines, $R-NH_2$, with nitrous acid has been extensively investigated in order to explain the proliferation of products which can arise from a single reaction. The most simple aliphatic primary amine, methylamine CH_3-NH_2, gives rise to several substitution products CH_3-X, where $X = -OH$, $-ONO$, $-NO_2$, $-Cl$, when treated with sodium nitrite and hydrochloric acid. On the other hand, in the aromatic series, aniline $C_6H_5-NH_2$, yields initially the ionic diazonium salt $C_6H_5-N_2^\oplus Cl^\ominus$ which can further react, as we will see, in a number of ways to give other aromatic derivatives.

4

KINETIC DATA

The kinetic rate expression for the reaction, which is independent of the products finally formed, has been found to be of the form

Rate α [amine] $[HNO_2]^2$

Thus, although an overall balanced equation for the production of methanol from methylamine would be

$$CH_3NH_2 + HNO_2 \rightarrow CH_3OH + N_2 + H_2O$$

that is, only involving *one* molecule of nitrous acid, the rate expression indicates that *two* such molecules are involved in the rate-determining stage of the reaction. Furthermore, under conditions of low acidity using a large excess of amine, the kinetic expression simplifies to

Rate α $[HNO_2]^2$

It therefore appears that nitrous acid, as such, is not the species responsible for the initial interaction with the amine; rather, two molecules of nitrous acid (or their equilibria equivalents), react to form some other entity, or entities, which then attack the amine.

In hydrochloric acid solution, the complete rate expression is:

Rate α [amine] $[HNO_2]^2$ + [amine] $[HNO_2]$ [HCl]

This includes an extra term which indicates that HCl plays a part in the reaction, especially at high acidities. This rate expression shows that at high acidity, when the second term becomes important in determining the reaction rate, HCl is equivalent to one of the nitrous acid molecules.

REACTING SPECIES PRODUCED IN SOLUTION

It has been shown that several species O=N—X are present in the $NaNO_2$/HCl solution at concentrations which depend upon the acidity of the solution:

$$O=N-NO_2 \qquad O=N-Cl \qquad O=N-\overset{\oplus}{O}H_2 \qquad O=\overset{..}{N}{}^{\oplus}$$
$$\text{(I)} \qquad\qquad \text{(II)} \qquad\qquad \text{(III)} \qquad\qquad \text{(IV)}$$

$\longrightarrow$
increasing acidity and concentration of species

These species are all derived from equilibria which involve two nitrous acid molecules (low acidity), or one nitrous acid molecule and its rate expression equivalent at higher acidities, HCl:

(I) 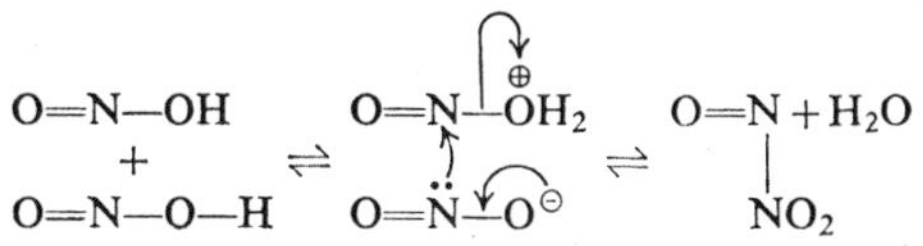 $O=N-OH$ $O=N-\overset{\oplus}{O}H_2$ $O=N+H_2O$
           + $\rightleftharpoons$ $\rightleftharpoons$ |
      $O=N-O-H$ $O=\ddot{N}-O^{\ominus}$ NO_2

                                                      nitrous anhydride

(II) 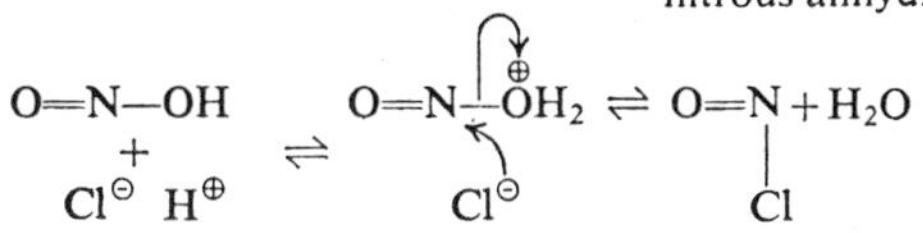 $O=N-OH$ $O=N-\overset{\oplus}{O}H_2$ $\rightleftharpoons$ $O=N+H_2O$
           + $\rightleftharpoons$ | |
      $Cl^{\ominus}$ $H^{\oplus}$ $Cl^{\ominus}$ Cl

                                                      nitrosyl chloride

(III) $O=N-OH+H^{\oplus} \rightleftharpoons O=N-\overset{\oplus}{O}H_2$ protonated nitrous acid

(IV) 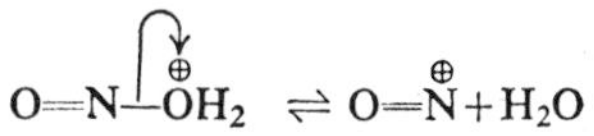 $O=N-\overset{\oplus}{O}H_2 \rightleftharpoons O=\overset{\oplus}{N}+H_2O$ nitrosonium ion

Thus, depending upon the precise reaction conditions, (I) to (IV) represent the possible attacking species. It will be seen that these are either electron deficient ((III) and (IV)), or potentially so by the presence of good leaving anionic groups ($NO_2^{\ominus}$ and $Cl^{\ominus}$ in (I) and (II)). They are therefore electrophiles which will seek out any available electron pair.

PRACTICAL PROCEDURE AND AMINE BASE STRENGTH

Amines possess one such pair of electrons on the nitrogen atom. In acidic solutions however this electron pair will tend to be used for proton coordination, for amines (substituted ammonias) are weak bases.

$$R-\ddot{N}H_2+H^{\oplus} \rightleftharpoons R-\overset{\oplus}{N}H_3$$

Here we encounter a most important practical detail for carrying out the reaction of nitrous acid with a primary aliphatic amine successfully. It is vital in these cases to have the reaction solution only faintly acidic, for any significant excess of acid would cause the concentration of the free amine, and hence the availability of the lone electron pair, to be almost zero. This is realized in practice by running the acid slowly into a solution of the amine and sodium nitrite in water. The nitrous acid is therefore generated *in situ* and the reaction proceeds to completion at room temperature, N_2O_3 and NOCl being the most likely attacking entities.

The aromatic primary amines on the other hand, are much weaker bases than their aliphatic counterparts due to the participation of the nitrogen lone pair in the resonance of the aromatic ring (see question (3), page 39). Base strengths are conveniently estimated in an analogous manner to acids by a pK_b value; $pK_b = -\log_{10}$ (base hydrolysis constant); the smaller the numerical value of pK_b the stronger the base to which it refers. Methylamine has a pK_b value of 3·4 compared with that for aniline of 9·4. The aromatic amines can therefore tolerate a much higher acidity in the nitrous acid reaction before the concentration of the unprotonated amine becomes negligible. In addition, in the presence of a significant excess of mineral acid, the reaction is likely to be facilitated in these cases because the more powerful electrophiles $H_2O^{\oplus}NO$ and $NO^{\oplus}$ can then operate. Consequently the nitrous acid reaction in the aromatic series is carried out in moderately concentrated acidic media.

THE INITIAL REACTION PATH

We are now in a position to formulate the initial path of the reaction which is of the same type for both aliphatic and aromatic primary amines, though the precise attacking electrophile NO.X may be different.

The initial and rate-controlling step is one of electrophilic attack by NO.X (where $X = NO_2$, Cl, $\overset{\oplus}{O}H_2$ or merely $\oplus$, as determined by the experimental conditions) on the lone electron pair on the nitrogen atom of the free amine, to yield an N-nitroso alkylammonium cation:

$$
\begin{array}{c}
\text{H} \\
| \\
\text{R--N:} \\
| \\
\text{H}
\end{array}
\quad
\begin{array}{c}
\text{O} \\
\| \\
\text{N--X} \\
\end{array}
\quad \xrightarrow{\text{slow}} \quad
\begin{array}{c}
\text{H} \\
| \\
\text{R--N}^{\oplus}\text{--N}{=}\text{O} + \text{X}^{\ominus} \\
| \\
\text{H}
\end{array}
$$

Alternatively of course, we can picture this step as one of nucleophilic attack by the amine on NO.X.

A series of relatively fast steps now occurs to yield a diazonium cation $R{-}N_2^{\oplus}$:

$$
\begin{array}{c}
\text{H} \\
| \\
\text{R--N}^{\oplus}\text{--N}{=}\text{O} \\
| \\
\text{H}
\end{array}
\quad \xrightarrow[\text{deprotonation}]{\text{fast}} \quad
\begin{array}{c}
\text{H} \\
| \\
\text{R--N--N}{=}\text{O} \quad \text{H}^{\oplus}\\
\text{N-nitroso compound}
\end{array}
$$

rapid proton transfer

$$
\begin{array}{c}
\text{R--}\ddot{\text{N}}{=}\overset{\oplus}{\ddot{\text{N}}} + \text{OH}^{\ominus} \\
\text{the diazonium} \\
\text{cation}
\end{array}
\quad \xleftarrow[\text{ionization}]{\text{fast}} \quad
\begin{array}{c}
\text{R--N}{=}\text{N--O--H} \\
\text{the diazo hydroxide}
\end{array}
$$

SUBSEQUENT REACTIONS IN THE ALIPHATIC SERIES

The diazonium cation from an aliphatic amine has two resonance forms. For the methylamine derivative we have

$$CH_3-\overset{..}{N}\overset{..}{=}\overset{..}{N}{}^{\oplus} \leftrightarrow CH_3-\overset{\oplus}{N}\equiv\overset{..}{N}$$

The unit positive charge thus has but two close sites where it can be accommodated. Aliphatic diazonium cations are therefore unstable and spontaneously decompose to evolve the very stable nitrogen molecule, for example:

$$CH_3-\overset{\oplus}{N_2} \xrightarrow{\text{fast}} CH_3{}^{\oplus} + N_2 \uparrow$$

This sequence produces highly reactive electron deficient carbonium ions. The methyl carbonium ion produced from methylamine will, for example, react with any available electron pair donor in solution to yield the observed proliferation of products:

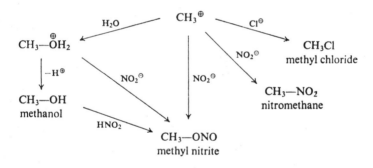

Methanol is produced by interaction of the carbonium ion with water, followed by proton loss to the solvent. Methanol can further react with nitrous acid to form methyl nitrite via protonated methanol:

$$O{=}N{-}O{-}H + CH_3{-}OH \rightarrow O{=}N{-}\overset{\ominus}{O} + CH_3{-}\overset{\oplus}{O}H_2$$

$$\downarrow$$

$$CH_3{-}ONO + H_2O$$

Methyl nitrite is also formed by direct interaction of the carbonium ion with nitrite anion. Nitromethane is formed by a similar interaction in which nitrogen becomes attached to carbon

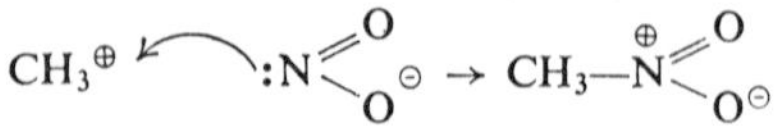

$$CH_3^{\oplus} \quad :N \overset{O}{\underset{O_{\ominus}}{<}} \rightarrow CH_3 - \overset{\oplus}{N} \overset{O}{\underset{O_{\ominus}}{<}}$$

and methyl chloride is the result of $CH_3^{\oplus}$ uniting with $Cl^{\ominus}$.

The precise proportions of the various products actually obtained depends enormously upon the conditions of reaction. For example, the yield of methanol from methylamine can vary from about 40 per cent to below 5 per cent for comparatively small differences in mineral acid concentration. A low yield of methanol is concurrent with a correspondingly higher yield of methyl nitrite.

With higher aliphatic amines, still further reactions are possible for the carbonium ion. It can stabilize itself by losing a proton to form an olefin, for example:

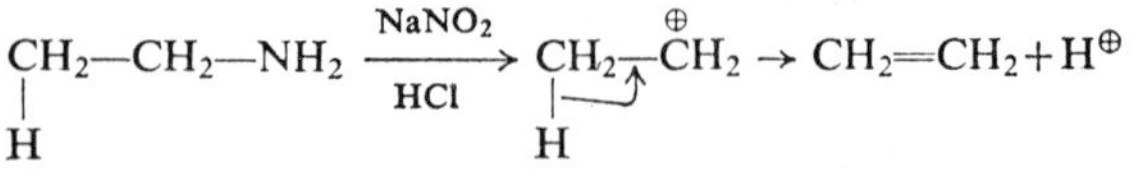

$$CH_2-CH_2-NH_2 \xrightarrow[HCl]{NaNO_2} CH_2{\overset{\oplus}{-}}CH_2 \rightarrow CH_2{=}CH_2 + H^{\oplus}$$
$$| |$$
$$H H$$

ethylamine   ethylene

or it can undergo a rearrangement reaction (see page 97).

SUBSEQUENT REACTIONS IN THE AROMATIC SERIES

In contrast to their aliphatic counterparts, aromatic diazonium cations possess a certain degree of stability. This is because the unit positive charge, which is formally located on nitrogen, can enlist the effective co-operation of the aromatic ring as an electron source, and the ion thus gains an additional stabilization by delocalization of the positive charge in the ring.

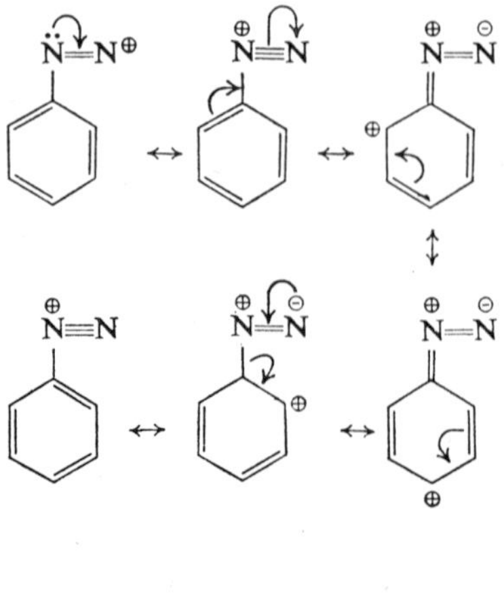

The summary structure for the benzene diazonium cation, produced from aniline by aqueous nitrous acid by 5°C, is

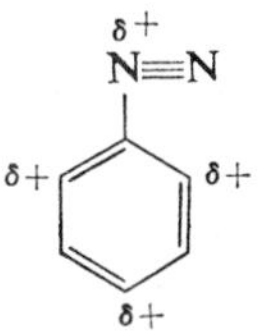

In hydrochloric acid solution, this exists formally as ions from the ionic salt, benzene diazonium chloride $C_6H_5\overset{\oplus}{-}N_2\overset{\ominus}{Cl}$. This colourless salt is difficult to isolate as it is very unstable when dry. Thus diazonium salts are normally prepared and stored in aqueous solution at 0–5°C.

If the aqueous solution of benzene diazonium chloride is warmed to about 70°C, phenol is produced.

$$C_6H_5-N_2{}^{\oplus}Cl^{\ominus} + H_2O \rightarrow C_6H_5-OH + N_2 + H^{\oplus} + Cl^{\ominus}$$

The diazonium cation is clearly unstable under these conditions and it first sheds its molecule of nitrogen to leave a positively charged phenyl carbonium ion.

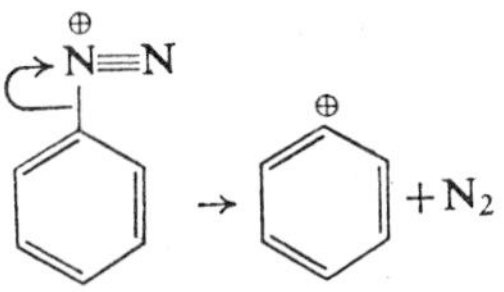

In contrast to the majority of aromatic species, this entity is distinctly electron deficient and is readily attacked by nucleophiles. In the aqueous solution, water will be the most predominant nucleophile and will react thus:

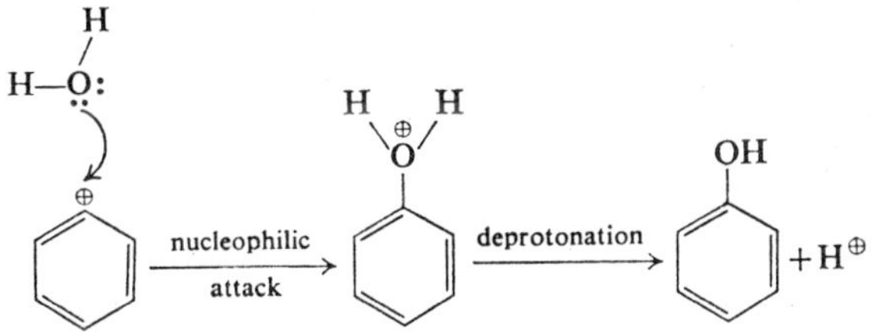

Questions

1 Give resonance structures for the nitrite anion.
2 At low acidity, what term could you substitute for $[HNO_2]^2$ in the rate expression

Rate α [amine] $[HNO_2]^2$?

48

3 Will ortho-nitroaniline be a stronger or weaker base than aniline? Give your reason.
4 Predict what would be the most probable major product from the reaction of methylamine with nitrous acid using methanol as solvent. Show the final stages of the reaction.
5 Predict the product formed when an aqueous solution of benzene diazonium chloride is treated with warm potassium iodide solution. What type of reaction is this?

B. Secondary amines

When secondary amines R—NH—R' (R, R' = aliphatic or aromatic) react with nitrous acid, no nitrogen is evolved. Instead, yellow oils or precipitates form which are identified as the N-nitrosamines.

$$\begin{array}{c} R \\ \diagdown \\ \diagup NH + HONO \rightarrow \\ R' \end{array} \begin{array}{c} R \\ \diagdown \\ \diagup N-N{=}O + H_2O \\ R' \end{array}$$

These compounds are formed by a similar pathway to that shown previously for primary amines, viz:

$$R'-\underset{\underset{H}{|}}{\overset{\overset{R}{|}}{N}}: \overset{O}{\underset{}{\overset{\|}{N}}}-X \rightarrow R'-\underset{\underset{H}{|}}{\overset{\overset{R}{|}}{N}}{}^{\oplus}-N{=}O \rightarrow R'-\underset{}{\overset{\overset{R}{|}}{N}}-N{=}O + H^{\oplus}$$
$$+ X^{\ominus}$$

The N-nitrosamine cannot undergo further reaction as for primary amines because it does not contain an N—H bond, that is, the next step, which would be one of proton transfer, cannot occur.

Aromatic nitrosamines can undergo further acid catalysed reactions, but these are too complex to discuss here.

C. Tertiary amines

Again with tertiary amines, no nitrogen is evolved on treatment with nitrous acid. A colourless solution of the amine nitrite salt is produced in the aliphatic series from which the amine can be reliberated by the addition of a strong base.

$$R'-\underset{\underset{R''}{|}}{\overset{\overset{R}{|}}{N}}: + HNO_2 \rightarrow R'-\underset{\underset{R''}{|}}{\overset{\overset{R}{|}}{N}}{}^{\oplus}-H \quad NO_2{}^{\ominus}$$

amine nitrite salt

In the aromatic series, a more complex reaction takes place which involves nitrosation (substitution by the group —N=O) of the aromatic ring by an electrophilic entity of the NO.X type. The —N̈(CH₃)₂ group of dimethyl aniline is an electron donor to the aromatic ring, and is therefore an activating group and ortho-para directing. We can represent C-nitrosation by 'nitrous acid' on this molecule at the para position thus:

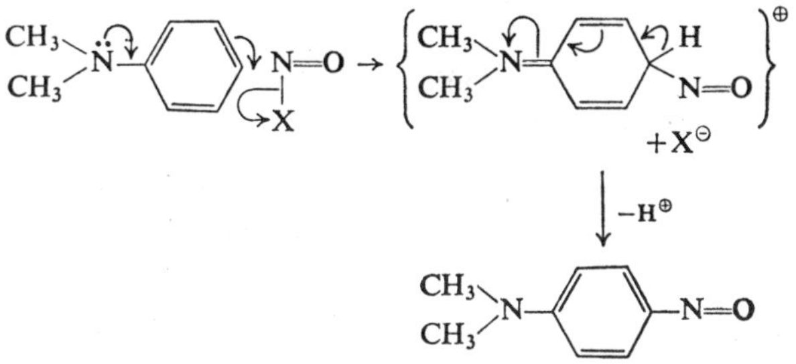

Again we see the familiar two stage process of electrophilic aromatic substitution; electrophilic attack at a position of high electron density, followed by deprotonation. The product from the reaction, p.nitrosodimethylaniline exists in the $NaNO_2/HCl$ reaction medium as the yellow hydrochloride salt, and it is rapidly formed even at 0°C. The free nitrosoamine, which is a green solid, is liberated from the reaction solution by the addition of alkali.

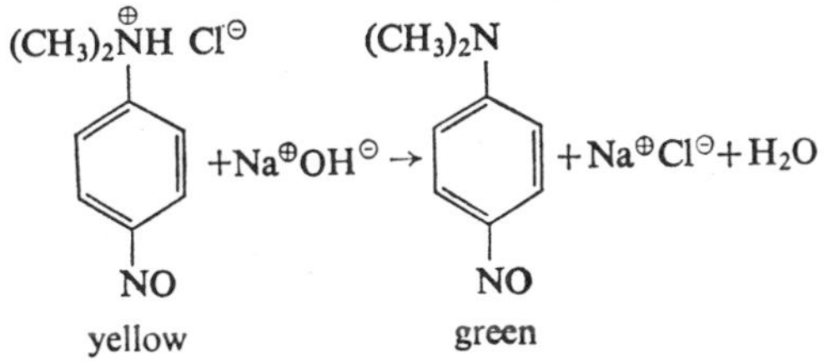

The varied nature of the reaction of nitrous acid with amines is often used to distinguish, on a qualitative basis, between different types of amine.

SUMMARY

In this section we have seen that in order to obtain a good yield of the desired product in a reaction, conditions must be selected which have taken into account both the mechanism of the reaction and the particular structure of the substrate. We have seen that carbonium ions are highly reactive, and that when they occur in the aromatic series the less common process of nucleophilic aromatic substitution takes place.

2.7 The coupling reaction of benzene diazonium chloride

The aromatic diazonium cation is an electrophilic entity. It is not surprising therefore that it can participate as the attacking species in electrophilic aromatic substitution reactions. $C_6H_5—\overset{\oplus}{N_2}$ is however not so strongly electrophilic as $NO_2{}^\oplus$, $Br^\oplus$ and $CH_3{}^\oplus$, and it will not therefore react with unactivated or deactivated aromatic nuclei. Its most common reactions are with phenols and aromatic amines, —OH and —NR$_2$ (R=H or alkyl) being activating groups. Para substitution by the diazonium cation predominates considerably over ortho substitution, largely because the cation is a very bulky incoming group, and it would tend to clash with the substituent already present in the aromatic ring if it entered at the ortho position. No such spatial congestion arises at the para position where electrons are made available in an efficient manner by the long range resonance effect. The reaction of benzene diazonium chloride with phenol (a 'coupling' reaction, though in fact a normal electrophilic aromatic substitution) can be represented thus:

p. hydroxyazobenzene (orange solid)

This reaction takes place rapidly at about 5°C at an intermediate pH. The cold solution of benzene diazonium chloride in hydrochloric acid is slowly stirred into a chilled alkaline solution of phenol. A bright orange precipitate is spontaneously formed from the colourless solutions. An alkaline solution of phenol is used because this contains the more strongly activated phenate ion $C_6H_5—\overset{..}{\underset{..}{O}}:{}^\ominus$, which will be a more effective nucleophile than phenol for the reaction; nucleophilic response at the para position will be more marked.

Reactions of diazonium cations with amines frequently occur by a similar process. Again an intermediate pH is employed in which the two reacting entities, the diazonium cation (not stable at high pH) and the *free* amine (protonated at low pH), can have a satisfactory existence. This forms the basis of an additional reason for preparing aromatic diazonium compounds from their primary amines with nitrous acid in moderately concentrated acid solution (see page 44); self-coupling reactions between the diazonium cation formed and any residual free amine are minimized, because the latter will be protonated, and hence the activating effect of the $-\overset{..}{N}H_2$ group destroyed, in the significantly acidic medium.

The coupling reaction is of importance in the preparation of many organic dyestuffs and indicators. Many varied colours can be obtained by suitable variation of the diazonium compound and/or the phenol or amine. For example, the well known indicator methyl orange is prepared by the sequence

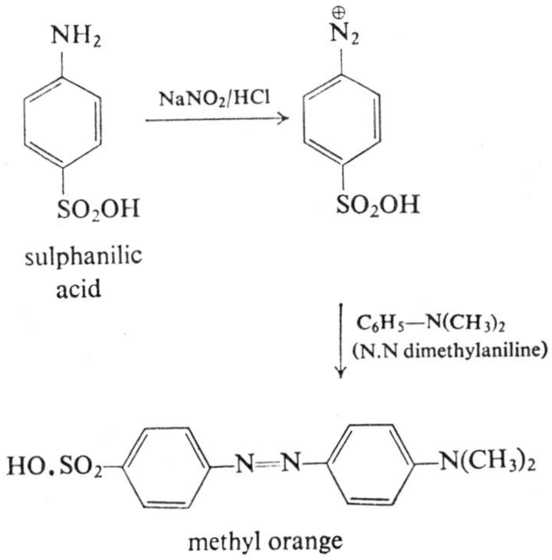

sulphanilic
acid

methyl orange

Question

Write down the formulae of products formed (if any) when an aqueous solution of benzene diazonium chloride reacts with (a) benzene (b) N.N diethylaniline (c) nitrobenzene (d) ortho-cresol (= ortho-methyl phenol), at room temperature.

3   Addition Reactions

3.1   Of olefins

A. *The bromination of ethylene*

$$CH_2{=}CH_2 + Br_2 \rightarrow \underset{\overset{|}{Br}}{CH_2}{-}\underset{\overset{|}{Br}}{CH_2}$$

1,2-dibromoethane (b.p. 112°C)
(ethylene dibromide)

This is an example of the saturation of an olefinic bond with free halogen. When gas jars of bromine and ethylene are allowed to mix at room temperature, a colourless oily liquid is produced.

The two electron pair bonds which join the carbon atoms of ethylene are not identical. One is a normal σ bond, analogous to the carbon–carbon bond of ethane, in which the electron distribution is concentrated along the line joining the nuclei of the two atoms. The other is a π bond, and the electron pair of this bond lies in an orbital above and below the plane of the atomic nuclei.

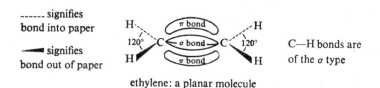

ethylene: a planar molecule

These bonds are the result of the sp² hybridization of carbon which leads to three trigonal σ bonds with bond angles of 120° (planar) and one π bond. The latter is formed by the overlap of the $2p_z$ electrons of carbon in its valence state. The important factor with regard to the reactivity of ethylene and other olefins is that the π electrons are less firmly held by the nuclei than the σ electrons. The π orbital is, like the π electron cloud of benzene, readily accessible and will attract electron deficient (electrophilic) entities; it will also tend to repel electron rich (nucleophilic) entities. The π orbital is therefore a potential electron source and a shield against nucleophilic attack.

Reaction of ethylene with bromine can be considered to occur thus: Under the influence of the π electron cloud, the approaching halogen molecule becomes polarized (electrically distorted). The π orbital then attracts the electron deficient end of the polarized bromine molecule, and by a concerted shift of electrons a positive bromonium ion and a negative bromide ion are formed. This represents stage one of the reaction.

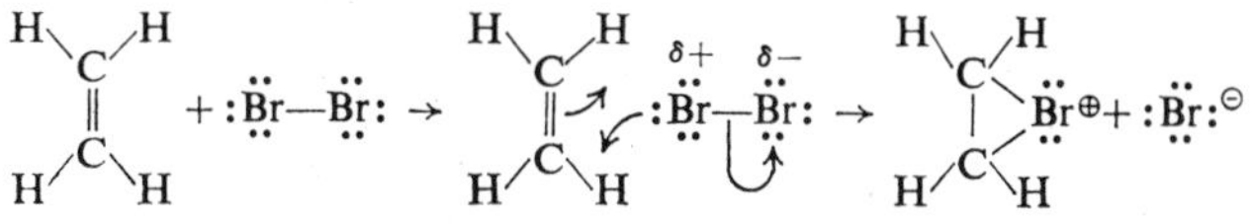

Stage two of the reaction then involves attack by the negative bromide ion from the rear on the intermediate bromonium ion:

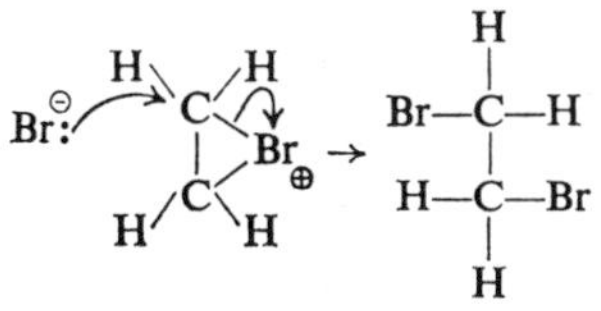

Rear attack by bromide ion is most probable because the bulky bromine atom in the bromonium ion hinders approach from the front.

In this reaction we see the halogen molecule acting initially as an electrophilic reagent by splitting out $Br^{\ominus}$ from the polarized molecule and reacting effectively as $Br^{\oplus}$. Alternatively we may view the double bond as a nucleophilic reagent displacing a bromide anion from each bromine molecule.

The two stage mechanism explains why *two* products are formed when ethylene reacts with bromine water. The intermediate bromonium ion can react not only with $Br^{\ominus}$ but also with water molecules to yield the bromohydrin:

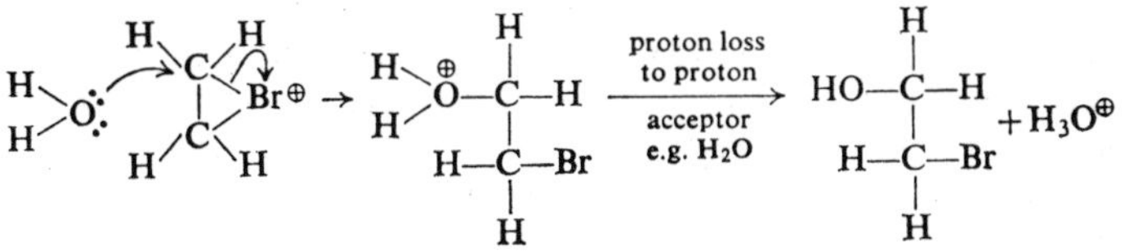

ethylene
bromohydrin

The second stage of rear attack has stereochemical implications which can be examined in certain cases. For example, addition of bromine to maleic and fumaric acids (geometrical isomers) gives rise to racemic and meso 2,3 dibromosuccinic acids respectively:

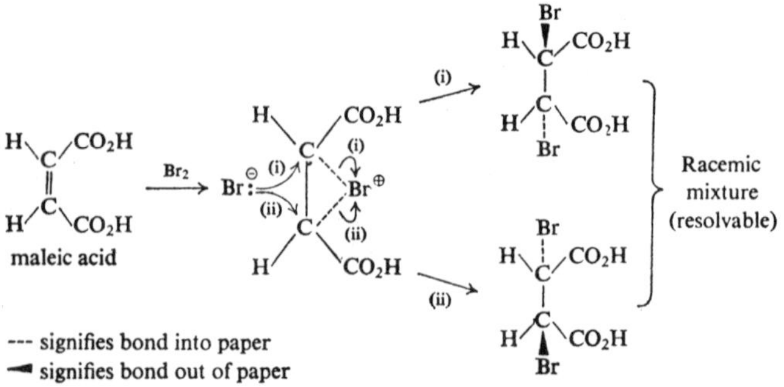

--- signifies bond into paper
◄ signifies bond out of paper

Racemic mixture (resolvable)

There is an equal probability of bromide anion attack from the rear (above the plane of the paper as depicted in the formulae) by either path (i) or path (ii) to yield equal quantities of two dibromo compounds which are non-superimposable mirror images, that is, a resolvable externally compensated racemic mixture. The two analogous paths for fumaric acid are:

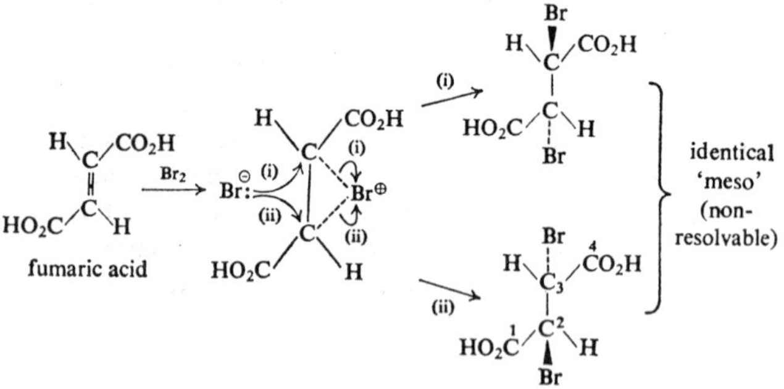

identical 'meso' (non-resolvable)

Rotation of the lower formula through 180° in the plane of the paper shows that it is superimposable upon the upper formula. Hence both paths in this case yield the same product. Rotation of the lower portion of either formula about the main C_2—C_3 bond through 180° yields

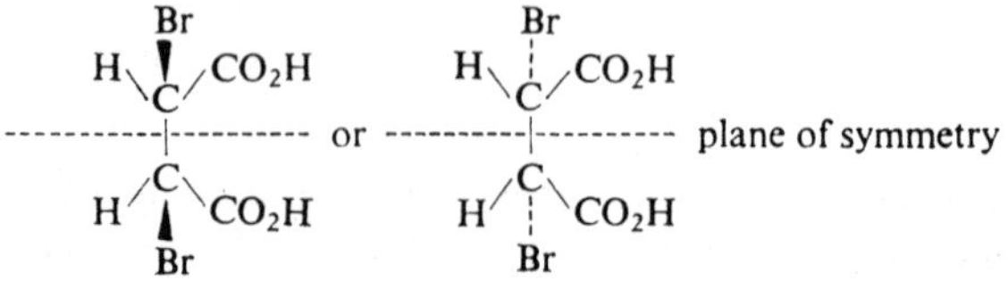

or ———— plane of symmetry

A plane of symmetry can be readily seen in these formulae and the product of bromination of fumaric acid is the internally compensated meso dibromide. The stereochemistry of the 2,3 dibromosuccinic acids is entirely analogous to that of the tartaric acids, which possess —OH instead of —Br. These results confirm that the addition of halogens to olefins is a two stage process in which the two atoms enter the molecule on opposite sides, that is, the reaction is a *trans* addition process.

Questions

1 Write down the formulae of the three products which are formed by reaction of ethylene with bromine water containing dissolved sodium chloride.

2 Both *cis* and *trans* 2-butene yield optically inactive products upon bromination. Name the isomer which yields a product which can be separated into two optically active forms.

B. *Catalytic hydrogenation*

Olefins, when treated with molecular hydrogen in the presence of a suitable catalyst, become saturated. Ethylene for example yields ethane:

$$CH_2{=}CH_2 + H_2 \xrightarrow{\text{catalyst}} \underset{\underset{H}{|}\quad\underset{H}{|}}{CH_2{-}CH_2}$$

Finely divided metals, which have the capacity of absorbing fairly large quantities of hydrogen, are the best catalysts. The most important are nickel, platinum and palladium, and these act as a reservoir for hydrogen. This reservoir is utilized when the olefin comes into contact with the catalyst surface.

The most usual procedure for catalytic hydrogenation is to shake the olefin (usually dissolved in an inert solvent such as ethanol or ethyl acetate) in an atmosphere of hydrogen, together with the finely divided catalyst which may be deposited on a charcoal support. The reaction normally proceeds effectively at room temperature and pressure, and the uptake of hydrogen can be followed volumetrically, each centre of unsaturation requiring one mole of hydrogen per mole of compound.

In contrast to the addition of bromine to olefins, hydrogen added in this way has been shown to do so by a *cis* addition process; that is, both hydrogen atoms add on to the olefin from the same side of the molecule. The sequence of the reaction is therefore most simply explained in the following terms.

56

Hydrogen molecules approach the catalyst surface where they are probably absorbed into the body of the catalyst as reactive free atoms. The olefin molecule then approaches the catalytic surface and forms an intermediate addition complex with the catalyst which involves the π electrons of the olefin. Transfer of hydrogen atoms from the catalyst on to the same side of the olefinic bond now takes place, and the reduced product is then readily desorbed to leave the catalyst surface free for further reaction.

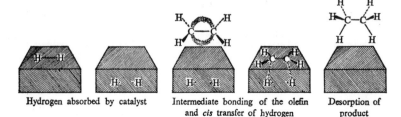

Hydrogen absorbed by catalyst | Intermediate bonding of the olefin and *cis* transfer of hydrogen | Desorption of product

Questions

1 How many moles of hydrogen will one mole of vitamin A of formula

absorb upon catalytic hydrogenation?

2 Acetylenes upon hydrogenation normally absorb two molecules of hydrogen per molecule of acetylene. Name and give the formula of the product obtained when methyl acetylene, CH_3—$C\equiv CH$, is so treated.

3 By careful choice of reaction conditions, usually employing a partially poisoned catalyst, acetylenes can be hydrogenated with one mole of hydrogen per mole of acetylene to yield an intermediate olefin. Name and give the formula of the product produced when dimethyl acetylene, CH_3—$C\equiv C$—CH_3, is so treated.

C. *Hydrobromination: Markovnikov's rule*

When olefins are treated with hydrogen halides either in the gas phase, or in an inert nonionizing solvent (for example, pentane), addition to form alkyl halides takes place. Hydrogen bromide reacts with ethylene to yield ethyl bromide:

$$CH_2{=}CH_2 + H{-}Br \rightarrow CH_3{-}CH_2{-}Br$$

With the next higher homologue, propylene, two modes of reaction are possible depending upon which way the hydrogen bromide molecule adds on to the olefinic bond:

(i) $\overset{3}{C}H_3{-}\overset{2}{C}H{=}\overset{1}{C}H_2 + H{-}Br \rightarrow CH_3{-}CH_2{-}CH_2{-}Br$
        1-bromopropane

or (ii) $CH_3{-}CH{=}CH_2 + Br{-}H \rightarrow CH_3{-}\underset{\underset{Br}{|}}{C}H{-}CH_3$

2-bromopropane

Hydrogen bromide has a permanent dipole moment (0·8 D) due to the greater electronegativity of bromine compared with hydrogen (see table, page 9). The molecule therefore has a permanent polarization

$$\overset{\delta+}{H}{-}\overset{\delta-}{Br}$$

Due to the electronic structure of olefins, notably the accessibility of the π electrons, the initial step of hydrogen halide addition, like that of bromination discussed in section A, must involve electrophilic attack. As the hydrogen bromide molecule approaches the π electron cloud of the olefin, its permanent polarization will be enhanced, with the $\delta+$ hydrogen atom of the hydrogen bromide closer to the π electron source than the bromine atom. In this position, the olefin molecule will also be polarized by the close hydrogen bromide molecule, and the first stage of addition to ethylene can be pictured:

$$\begin{array}{ccc} \delta+CH_2 & & \overset{\oplus}{C}H_2 \\ \| \quad \searrow \quad {}^{\delta+} \quad {}^{\delta-} \rightarrow & | & +Br^{\ominus} \\ \delta-CH_2 \downarrow \quad H{-}Br & CH_2{-}H & \end{array}$$

This is essentially electrophilic attack by $H^{\oplus}$ on the electron rich olefin, and it will naturally be followed by rapid combination of the carbonium ion with the bromide anion:

$$H{-}CH_2{-}\overset{\oplus}{C}H_2 + {:}\overset{..}{\underset{..}{Br}}{:}^{\ominus} \xrightarrow{\text{fast}} H{-}CH_2{-}CH_2{-}Br$$

These steps are very similar to those described for the bromination of ethylene.

5

For propylene, the mechanism of hydrobromination must be analogous to that for ethylene. On a simple basis, we would expect that there would be an equal probability of reaction proceeding by path (i) or path (ii) to give equal yields of 1 and 2-bromopropane. In practice, however, under normal conditions, the reaction proceeds almost exclusively by route (ii) to give 2-bromopropane as product.

The added hydrogen atom clearly prefers to become attached to the 'end' carbon atom of the olefinic bond. As this atom becomes attached by electrophilic attack as $H^{\oplus}$, it will seek out the centre of highest electron density. We must conclude therefore that the propylene molecule is polarized, or becomes polarized as the reagent approaches, with the end carbon atom (C_1) the most negative.

$$\underset{3}{CH_3}-\overset{\delta+}{\underset{2}{CH}}=\overset{\delta-}{\underset{1}{CH_2}}$$

ELECTRON REPULSION BY ALKYL GROUPS

It will be recalled that earlier (page 12) the decreased acidity of acetic and propionic acids compared with formic acid, was explained by the apparent ability of the alkyl groups to 'push' electrons on to the carboxyl group, thus discouraging deprotonation. A similar explanation is invoked in order to explain the increased basicity of alkyl substituted ammonias compared with ammonia itself.

Compound	Formula	pK$_b$ value	
ammonia	NH_3	4·75	
methylamine	CH_3NH_2	3·4	
ethylamine	$CH_3CH_2NH_2$	3·3	Stronger bases than
dimethylamine	$(CH_3)_2NH$	3·2	ammonia
diethylamine	$(CH_3CH_2)_2NH$	3·0	

The effect of the alkyl groups is clearly to make the unshared pair of electrons on nitrogen more available for the coordination of a proton than in ammonia, by an electron donating effect, for example:

$$CH_3 \longrightarrow \overset{\displaystyle H}{\underset{\displaystyle H}{N:}}$$

Again, the $-CH_3$ group was given (page 40) as an activating, ortho-para directing substituent in aromatic compounds. Toluene, $C_6H_5-CH_3$, for example, is more readily nitrated than benzene and yields initially a mixture of ortho and para nitrotoluenes. Electron donation by the methyl group to the aromatic ring must therefore take place, and the toluene dipole moment of 0·4 D confirms this.

It is not easy to explain why alkyl groups should exhibit this electron donating effect, and although several explanations have been suggested, the matter must at present be considered to be unsettled. However the phenomenon undoubtedly exists, and it can help us to explain the normal mode of addition of hydrogen bromide and other reagents to propylene.

The methyl group of propylene will release electrons towards the adjacent carbon atom and give it a *temporary* partial negative charge.

$$CH_3 \rightarrow \overset{\delta-}{C}-CH_2$$

The charge is temporary, because once it is induced on the central carbon atom, it will tend to repel the more readily distortable π electrons, by an effect somewhat analogous to the shunting of railway trucks, to give the permanent dipole thus

$$CH_3 \rightarrow \overset{\delta+}{C}\underset{}{=\!=}\overset{\delta-}{C}H_2$$

the electron releasing effect of the methyl group now stabilizing the partial *positive* charge on the central carbon atom. This latter structure for propylene is in agreement with the small, but significant, observed dipole moment of 0·4 D.

THE PROPYLENE REACTION PATH

Having explained why carbon atom 1 of propylene is the most negative in the molecule, it is now possible to give the reaction sequence.

$$
\begin{array}{ccc}
CH_3 & & CH_3 & & CH_3 \\
| & \text{electrophilic} & | & \text{nucleophilic} & | \\
\delta+CH & \xrightarrow{\text{attack}} & {}^{\oplus}CH \;+\; :\!\overset{..}{Br}\!:^{\ominus} & \xrightarrow{\text{attack}} & CH-Br \\
\| \quad \delta+ \;\; \delta- & \text{by reagent} & | & \text{by bromide} & | \\
\delta-CH_2\,\,H-Br & & CH_3 & \text{anion} & CH_3 \\
& & \text{intermediate} & & \\
& & \text{secondary} & & \\
& & \text{carbonium ion} & &
\end{array}
$$

It should be noted that the intermediate carbonium ion is secondary in nature, and this is an additional factor which favours the observed mode of hydrogen bromide addition. The alternative pathway would yield $CH_3-CH_2-\overset{\oplus}{CH_2}$ as the intermediate carbonium ion. This is a primary carbonium ion, and it is less stable, and therefore less readily formed, than the secondary carbonium ion, because the positively charged carbon atom has one less carbon atom as an immediate neighbour over which it can distribute its unit positive charge by resonance. The positive charge of the secondary carbonium ion $CH_3\overset{\oplus}{C}HCH_3$ is more central in the total 'electron sea' of the entity. (Tertiary carbonium ions, such as $(CH_3)_3C^\oplus$, are similarly more stable than their secondary counterparts.)

MARKOVNIKOV'S RULE

The mode of addition of unsymmetrical reagents to olefins is summarized by Markovnikov's empirical generalization. This states that 'the ionic addition of unsymmetrical reagents to unsymmetrical olefins proceeds in such a way that the more positive part of the reagent becomes attached to the least substituted carbon atom of the double bond'. When the hydrogen halides or water are the added molecules, a hydrogen atom constitutes the 'more positive part', and the rule is then conveniently stated as 'to him that hath shall be given', or in other words, the hydrogen atom becomes attached to the olefinic carbon atom which already carries the larger number of hydrogen atoms, for example:

$$(CH_3)_2C{=}CH_2 + HBr \rightarrow (CH_3)_2\underset{\underset{Br}{|}}{C}{-}CH_3$$

$$(CH_3)_2C{=}CHCH_3 + HBr \rightarrow (CH_3)_2\underset{\underset{Br}{|}}{C}{-}CH_2CH_3$$

ANTI-MARKOVNIKOV ADDITION

It has been observed that the addition of hydrogen bromide to propylene in the presence of peroxides yields predominantly 1-bromopropane, that is, the reagent adds on to the olefin under these conditions in a manner contrary to Markovnikov's rule. This is due to a change in the mechanism of the reaction.

Peroxides are relatively unstable compounds, and are well known initiators of free radicals.

$$RO{-}OR \rightarrow 2RO\cdot \quad (R \text{ often} {=} C_6H_5{-}CO{-})$$

The radicals so produced can start a self propagating 'chain reaction' by encounter first with hydrogen bromide molecules to form bromine radicals,

$$RO\cdot + H{-}Br \rightarrow RO{-}H + Br\cdot$$

which then attack the propylene molecule to give (by analogy with carbonium ions), a more stable secondary (as opposed to primary) radical addition species:

$$CH_3{-}CH{=}CH_2 + Br\cdot \rightarrow CH_3{-}\overset{\cdot}{C}H{-}CH_2Br$$

As before, a 'hot object', in this case an odd unpaired electron, can be more readily held within a system if it is near the centre of the system rather than on the periphery. The secondary radical then reacts with another hydrogen bromide molecule to yield the product, and another bromine radical which can further propagate the reaction:

$$CH_3—\overset{\cdot}{C}H—CH_2Br + HBr \rightarrow CH_3—CH_2—CH_2Br + Br\cdot$$

It can be seen therefore that the change in the mode of addition of the reagent is due to a change from an ionic mechanism to a free radical mechanism. Markovnikov addition requires initiation by $H^{\oplus}$; Anti-Markovnikov addition requires initiation by $Br\cdot$. Each species attacks the olefin molecule at the centre of highest electron density to yield the most stable intermediate carbonium ion or free radical. Scrupulous purity of reagents is necessary in hydrogen bromide additions if mixtures of products resulting from both ionic and free radical addition are not to be obtained.

Questions

1 Acetylene will add two molecules of hydrogen bromide, proceeding through an intermediate olefinic compound. Write down the formula of the final product obtained under conditions selected so that oxygen was excluded from the reaction vessel. By what general type of mechanism does the reaction proceed?

2 Hypochlorous acid is polarized $\overset{\delta-}{HO}—\overset{\delta+}{Cl}$. Give the formula of the product resulting from the addition of hypochlorous acid to propylene.

3 Ionizing solvents, such as water, are not normally used for the addition of hydrogen halides to olefins because the solvent can participate in the reaction. Dilute aqueous solutions of halogen acids therefore yield significant quantities of alcohols from olefins by the net addition of the elements of water across the olefinic bond. Give clearly the reaction sequence when propylene is so treated to yield an alcohol. Give the name of the alcohol produced.

4 Hydrogen bromide can be split into free radicals by ultra-violet light of a specific wavelength.

$$H—Br \rightarrow H\cdot + Br\cdot$$

1-bromopropane is the major product, obtained by anti-Markovnikov addition, when propylene is treated with hydrogen bromide under these conditions. Give the reaction path, and suggest why an equal yield of 2-bromopropane is not obtained.

5 The addition of one molecule of hydrogen bromide to acetylene dicarboxylic acid is as follows

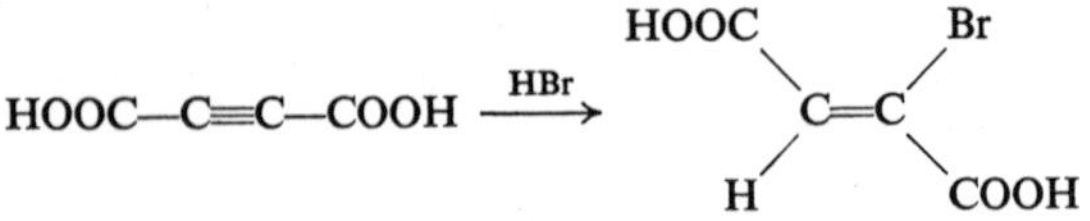

$$\text{HOOC—C}\equiv\text{C—COOH} \xrightarrow{\text{HBr}} \begin{array}{c} \text{HOOC} \\ \diagdown \\ \text{H} \diagup \end{array} \text{C}=\text{C} \begin{array}{c} \diagup \text{Br} \\ \diagdown \text{COOH} \end{array}$$

What does this tell you about the mode of addition of hydrogen halides
to multiple carbon–carbon bonds?

D. *Glycolisation*

The process

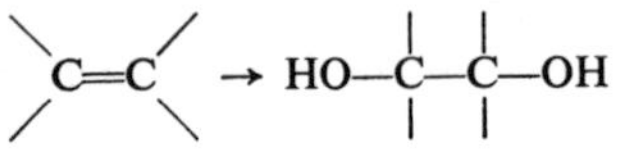

$$\diagup\text{C}=\text{C}\diagdown \;\rightarrow\; \text{HO—C—C—OH}$$

is known as a 1, 2 glycolization reaction; an —OH group is added on to each
carbon atom of a double bond. The most common reagents used for
effecting this transformation are potassium permanganate, osmium tetroxide (OsO_4), or organic peracids (R—CO—O—OH).

The first two reagents are known to react by a similar mechanism which
involves cyclic intermediates of the type

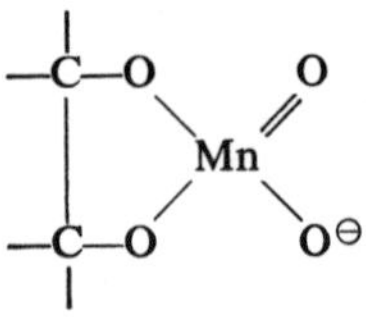

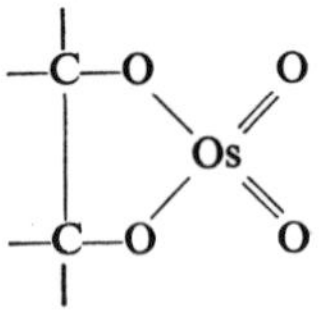

which are subsequently hydrolyzed in aqueous solutions to the glycol, and a
reduced form of the reagent. Using O^{18} labelled permanganate ion, it has
been shown that both oxygen atoms of the glycol are derived from the
oxygen atoms originally bound to manganese, and not from water added in
the hydrolysis stage. Hydrolysis of the intermediate cyclic osmic ester, which
can be isolated, can therefore, for example, be pictured

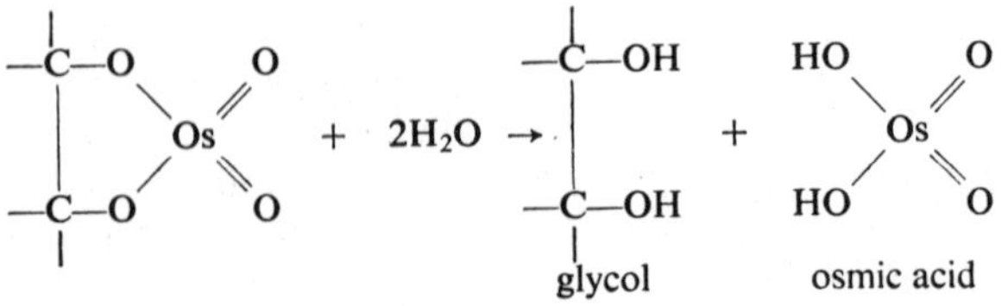

The permanganate reaction, which is usually carried out in cold, dilute, aqueous alkaline solution, is often difficult to control, for further oxidation of the glycol is liable to take place. Osmium tetroxide, though toxic and expensive, is rather more selective in its action. The osmic ester is usually formed in a dry inert solvent at 0°C. It is subsequently decomposed by the addition of an aqueous reducing agent, usually sodium bisulphite, which yields the glycol and reduced osmium compounds which are readily separated.

A consequence of the cyclic intermediates, formed by these reagents, is that the olefin molecule is 'held' by the cyclic structure, so that both hydroxyl groups are added to the olefin from the same side of the molecule. The process is therefore one of *cis*-addition.

Peracids on the other hand, glycolize olefins by a different mechanism. Peracids are generated by the action of hydrogen peroxide on carboxylic acids, for example:

$$CH_3-CO-O-H + HO-OH \rightarrow CH_3-CO-O-OH + H_2O$$

and are frequently used as a solution in ether or chloroform. They react initially with olefins as oxygen donors, to form an intermediate epoxide which can be isolated. Ethylene, for example, yields ethylene oxide:

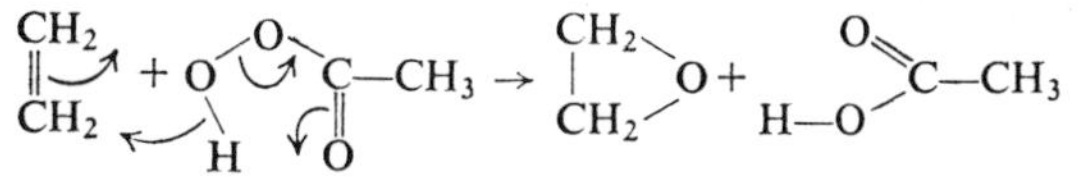

Subsequent aqueous acid hydrolysis of the epoxide yields the *trans*-glycol by 'rear' attack on the protonated epoxide by water:

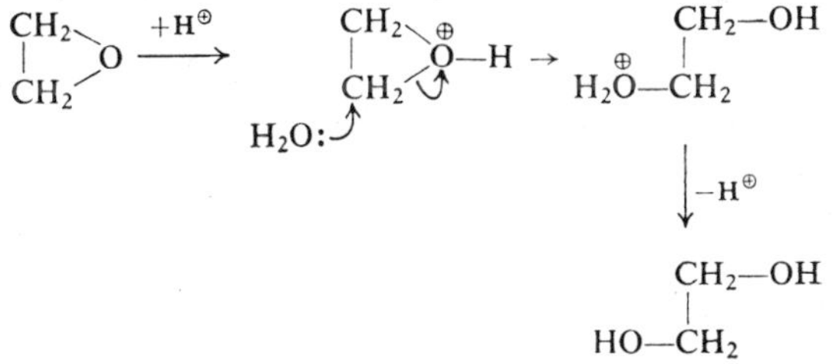

There is a close similarity between this oxide ring opening and the opening of the bromonium ion intermediate in the bromination of olefins (see page 53).

By careful choice of the reagent, the glycolization of olefins can therefore be stereospecifically controlled to yield either the *cis* or the *trans* 1, 2 glycol.

Questions

1 Is the statement, 'the glycolization of ethylene by potassium permanganate and peracetic acid gives rise to two different products, the *cis* and *trans* forms of ethylene glycol', true or false? Give your reason.

2 Name the glycolized products formed when maleic acid ($=cis$-ethylene dicarboxylic acid) is treated with (a) OsO_4 (b) CH_3CO_3H.

3 In the epoxidation of olefins in an ionizing solvent, peracetic acid is likely to ionize thus:

$$CH_3\text{—}COO\text{—}OH \rightleftharpoons CH_3COO^{\ominus} + {}^{\oplus}OH$$

Suggest a reason for this, and state which species will initially attack the olefin. Give the formula of a substance (apart from the glycol) which is likely to result from the opening of the ethylene oxide ring under these conditions.

E. *Ozonolysis*

Ozone undergoes ready reaction with olefins in an inert solvent at low temperatures to yield unstable addition compounds termed 'ozonides'. Ozone has a dipolar resonant structure

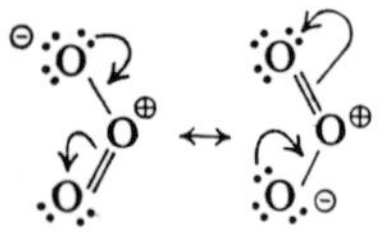

with a bond angle of about 120°. The central oxygen atom is relatively electron deficient, and can act as an electrophile for attack on the π electrons of the olefin. The final structure of the ozonide has been shown to have a structure resulting from complete rupture of the carbon–carbon double bond. The sequence by which this is formed is very complex, and we can conveniently show its formation thus

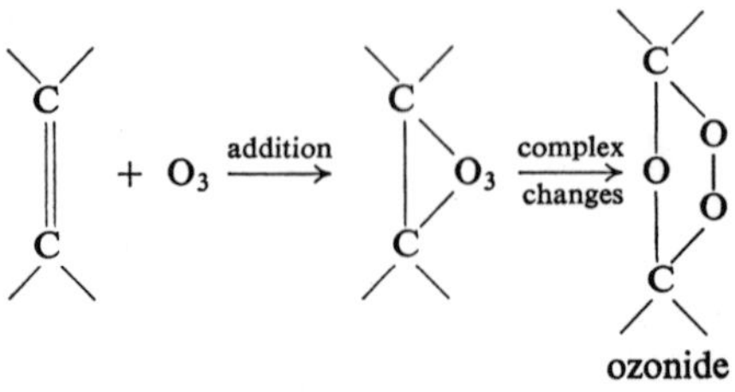

ozonide

The ozonides themselves are of little use, but on further treatment either with boiling water or hydrogen, they decompose to yield usually two fragments containing carbonyl groups:

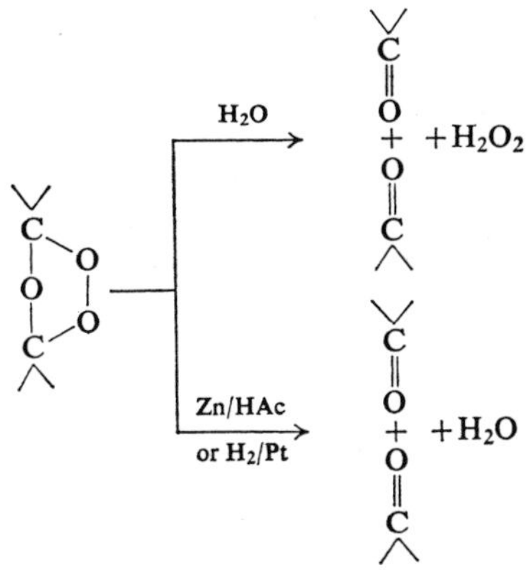

$$\begin{array}{c} \overset{\displaystyle\vee}{\underset{\parallel}{C}} \\ O \\ + \quad +H_2O_2 \\ O \\ \parallel \\ \underset{\displaystyle\wedge}{C} \end{array}$$

(with H_2O)

$$\begin{array}{c} \overset{\displaystyle\vee}{\underset{\parallel}{C}} \\ O \\ + \quad +H_2O \\ O \\ \parallel \\ \underset{\displaystyle\wedge}{C} \end{array}$$

(with Zn/HAc or H_2/Pt)

Thus, the overall reaction is replacement of $\diagup C = C \diagdown$ by two $\diagup C = O$ units, for example:

$$CH_3-CH=C\diagup^{CH_3}_{\diagdown CH_3} \quad \rightarrow \quad CH_3-CH=O + O=C\diagup^{CH_3}_{\diagdown CH_3}$$

acetaldehyde   acetone

Ozonolysis is therefore a very useful reaction for proof cf the structures of olefinic compounds.

Questions

1 Give the names of compounds produced when the following olefins are ozonized:

  (a) $CH_3CH=CHCH_3$
  (b) $(CH_3)_2C=CH_2$
  (c) $CH_3CH_2-C=CH_2$
       $\quad\quad\quad\quad |$
       $\quad\quad\quad CH_3$

2 What type of olefin will yield only one fragment on ozonolysis?

SUMMARY

In this section we have seen that the π electron cloud of olefins makes the first stage of addition of reagents to these compounds electrophilic in nature. The two added atoms or groups enter the molecule either on the same side (cis) or on opposite sides (trans) of the double bond.

Alkyl groups tend to repel electrons in a manner which is the reverse of the inductive pull by halogens. Markovnikov's rule predicts the position of addition of unsymmetrical reagents to unsymmetrical olefins.

Alkyl carbonium ions and radicals have the stability order:

Primary < secondary < tertiary

3.2 Of carbonyl compounds

COMPARATIVE REACTIVITY

Reference has been made (page 15) to the structure of acetone, which has a polarized carbon–oxygen double bond. The same is true for all aldehydes and ketones, the oxygen atom acquiring a greater share of the bonding electrons, notably the π electrons, than carbon, due to its greater electronegativity, that is,

$$\overset{\delta+}{>}\overset{\delta-}{C=O} \quad \text{or} \quad >C-O$$

It would therefore appear that carbonyl compounds should be susceptible to nucleophilic attack at the carbonyl carbon atom, and to electrophilic attack at the carbonyl oxygen atom. This is indeed found to be the case, though electrophilic attack on oxygen is only of significance where the electrophile is a proton.

However, as this type of reactivity of carbonyl compounds depends upon the carbonyl carbon atom being partially positive, the presence of neighbouring atoms or groups which have an electron donating effect towards this atom will lower the reactivity. We have seen (page 58) that alkyl groups tend to donate electrons. Hence the reactivity of formaldehyde will be greater than that of other aldehydes, which in turn will be greater than that of ketones.

$$\underset{O}{\overset{\|}{H-C-H}} > \underset{O}{\overset{\|}{R\rightarrow C-H}} > \underset{O}{\overset{\|}{R\rightarrow C \leftarrow R'}}$$

The reasons for the difference in reactivity between carboxylic acids and ketones was outlined earlier (page 23). When resonance effects can operate by lone pair donation towards a carbonyl carbon atom, thus outweighing any electronegativity induced pull away from this atom, then a marked decrease in reactivity of this carbon atom towards nucleophiles occurs. The following reactivity sequence is therefore observed:

$$R \rightarrow \underset{\underset{O}{\|}}{C} \leftarrow R' \gg R \rightarrow \underset{\underset{O}{\|}}{C} \rightarrow \ddot{O}R' >$$

$$R \rightarrow \underset{\underset{O}{\|}}{C} \rightarrow \ddot{N}H_2 > R \rightarrow \underset{\underset{O}{\|}}{C} \rightarrow \ddot{O} :^{\ominus}$$

Groups near to a carbonyl group which have an electron attracting effect conversely increase the susceptibility of the carbonyl carbon atom to attack by nucleophiles. For example, the addition hydrate of acetaldehyde cannot be isolated, whereas trichloroacetaldehyde forms a stable crystalline hydrate:

chloral hydrate

The inductive pull of the three chlorine atoms not only makes the carbonyl carbon atom more positive, but it also stabilizes the hydrate once formed, by preventing loss of —OH with its bonding electrons. (1,1-dihydroxy compounds are generally unstable and dehydrate to give the corresponding aldehyde.)

A. *Reduction by lithium aluminium hydride*

Lithium aluminium hydride, $LiAlH_4$, is formed in the reaction between lithium hydride and aluminium chloride in ether solution:

$$4LiH + AlCl_3 \rightarrow LiAlH_4 + 3LiCl$$

It is a grey-white solid, which is rapidly hydrolyzed by water with the evolution of hydrogen. It is moderately soluble in ether, and in this medium it is a most valuable reducing agent in organic chemistry. It reduces all carbonyl compounds (amides excepted) to the corresponding alcohols. Amides and nitriles are reduced to their corresponding amines, while olefinic and acetylenic bonds normally remain inert to the reagent. Reactions are often rapid and almost quantitative in many cases. Use of this reagent is often preferred over other reductive methods for carbonyl compounds (for example, H_2/Pt or Na/C_2H_5OH) which are often more difficult to carry out, or lead to undesirable side reactions.

The reagent furnishes aggregates of solvated lithium and aluminohydride ions, $Li^\oplus (AlH_4)^\ominus$, in the reaction medium. The complex $AlH_4^\ominus$ ion acts as a carrier of hydride ion, $H^\ominus$. Hydride ion is in turn a very powerful nucleophile towards the carbonyl carbon atom.

The reduction of acetone to isopropyl alcohol therefore proceeds as follows

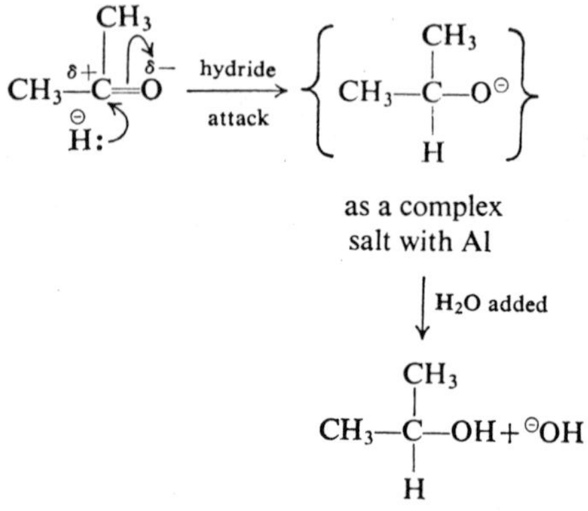

as a complex
salt with Al

H_2O added

One molecule of lithium aluminium hydride reduces four molecules of ketone, the net equation being

$$4(CH_3)_2C{=}O + Li^\oplus(AlH_4)^\ominus \rightarrow \{[(CH_3)_2CHO]_4Al\}^\ominus Li^\oplus$$

complex intermediate salt

$$\downarrow 4H_2O$$
$$\text{(hydrolysis)}$$

$$4(CH_3)_2CH(OH) + Al(OH)_3\downarrow + LiOH$$

Water, or preferably dilute sulphuric acid, is cautiously added at the end of the reaction to liberate the alcohol by hydrolysis of the intermediate complex salt. The alcohol is extracted into the ether layer leaving unwanted inorganic materials in the aqueous solution.

Questions

1 Give the formulae of the products formed when the compound $CH_3{-}CH{=}CH{-}CH_2{-}COOH$ is treated with (a) H_2/Pt (b) $LiAlH_4$.

2 Write down the formulae of the products produced from the reduction of (a) acetaldehyde, (b) methyl acetate, with $LiAlH_4$.

B. *Acetal formation*

In the presence of dry hydrogen chloride as catalyst, aldehydes will undergo reaction with alcohols to yield acetals. Acetaldehyde, for example, yields its dimethyl acetal with two molecules of methanol:

$$\underset{H}{\overset{CH_3}{>}}C=O + 2CH_3OH \underset{\text{dil.aq.HCl}}{\overset{\text{dry HCl}}{\rightleftharpoons}} \underset{H}{\overset{CH_3}{>}}C\underset{OCH_3}{\overset{OCH_3}{<}} + H_2O$$

Acetals are stable to alkali, and are therefore of use as a protective group for aldehydes, the parent aldehyde being regenerated by treating the acetal with dilute aqueous acid.

The mode of acetal formation principally involves proton equilibria and the ability of alcohols to act as nucleophiles (see page 7). The first step is protonation of the carbonyl oxygen atom.

$$\underset{H}{\overset{CH_3}{>}}\overset{\delta+}{C}=\overset{\delta-}{\ddot{O}} + \overset{\delta+}{H}-\overset{\delta-}{Cl} \rightarrow \underset{H}{\overset{CH_3}{>}}C=\overset{\oplus}{O}-H \leftrightarrow \underset{H}{\overset{CH_3}{>}}\overset{\oplus}{C}-O-H$$
$$+ Cl^{\ominus}$$

This enhances the electrophilic character of the carbonyl carbon atom which now undergoes nucleophilic attack by the alcohol to yield, after deprotonation, a hemi ('half') acetal:

$$\underset{\underset{H}{\overset{|}{O}}\diagdown CH_3}{\overset{CH_3}{>}\overset{\oplus}{\underset{H}{C}}\diagup O-H} \rightarrow \underset{H}{\overset{CH_3}{>}}C\underset{\overset{\oplus}{O}\diagdown CH_3}{\overset{O-H}{<}} \rightarrow \underset{H}{\overset{CH_3}{>}}C\underset{OCH_3}{\overset{OH}{<}} + H^{\oplus}$$

hemiacetal

Reprotonation of the hemiacetal at the —OH group followed by loss of water yields a reactive carbonium ion.

$$\underset{H}{\overset{CH_3}{>}}C\underset{OCH_3}{\overset{OH}{<}} + H^{\oplus} \rightarrow \underset{H}{\overset{CH_3}{>}}C\underset{OCH_3}{\overset{\overset{\oplus}{O}H_2}{<}} \rightarrow \underset{H}{\overset{CH_3}{>}}\overset{\oplus}{C}\underset{OCH_3}{} + H_2O$$

This entity rapidly reacts with another methanol molecule to yield, after deprotonation, the final acetal:

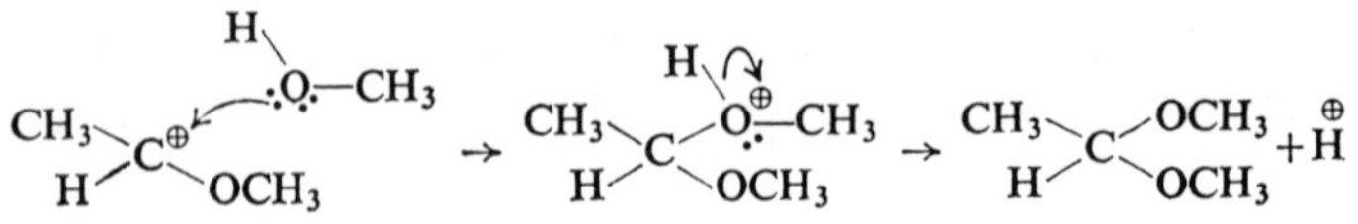

acetaldehyde
dimethyl acetal

All the individual steps in this reaction are reversible.

Question

1 Give diagrams showing the electron pair shifts for the aqueous acid hydrolysis of acetaldehyde hemiacetal.
2 Indicate how you would effect the following transformation, giving your reasons

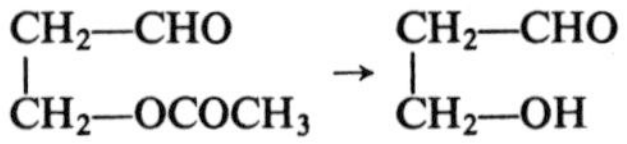

$$CH_2-CHO \atop CH_2-OCOCH_3 \quad \rightarrow \quad CH_2-CHO \atop CH_2-OH$$

C. Cyanohydrin formation

At a pH of about 4 (slightly acidic) hydrogen cyanide forms addition compounds with aldehydes and simple ketones. Acetone yields acetone cyanohydrin:

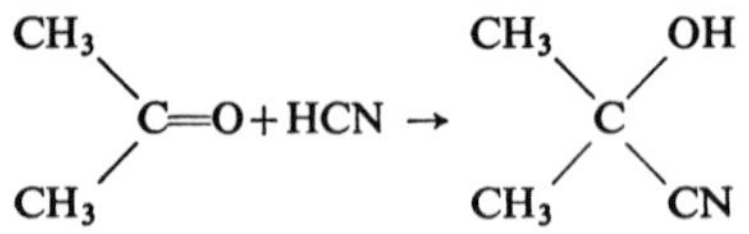

The reaction is very slow in the presence of excess acid, and a study of the addition has shown that the actual attacking species is $CN^{\ominus}$, the cyanide ion. The reaction therefore proceeds by nucleophilic attack by cyanide ion, followed by protonation in the slightly acidic medium:

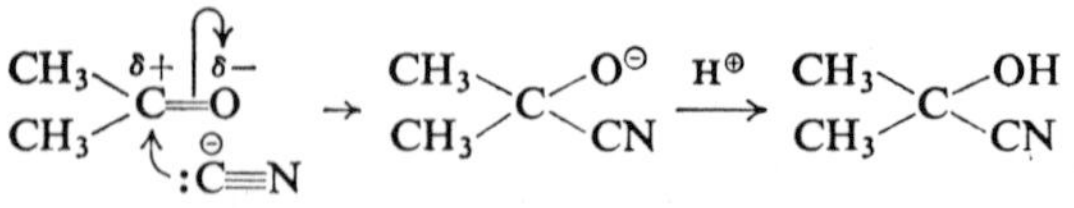

This is in contrast to the mode of acetal formation, where the first step is protonation of the oxygen atom of the carbonyl group. Although initial protonation of acetone in this reaction would undoubtedly render the carbonyl carbon atom more electrophilic,

$$\underset{CH_3}{\overset{CH_3}{\diagdown}}C{=}O + H^{\oplus} \;\rightleftharpoons\; \underset{CH_3}{\overset{CH_3}{\diagdown}}\overset{\oplus}{C}{\cdots}O{-}H$$

the low pH required to effect this step efficiently, would suppress the ionization of hydrogen cyanide (a weak acid):

$$HCN \;\rightleftharpoons\; H^{\oplus} + CN^{\ominus}$$

Thus at high acidity, the concentration of the necessary nucleophile, $CN^{\ominus}$, is low. In order to allow both steps of the reaction to proceed efficiently, an intermediate pH is selected. This is normally obtained by running a specific amount of mineral acid into an aqueous solution of the aldehyde or ketone containing dissolved sodium or potassium cyanide.

Questions

1 Acetaldehyde, like other aldehydes and methyl ketones, will form in aqueous solution a bisulphite addition compound with sodium bisulphite (acting as $Na^{\oplus\ominus}:SO_3H$). Show the reaction sequence.

2 Acetaldehyde yields an unstable addition product when treated with ammonia. Show the reaction path.

3 Why does ethyl acetate fail to yield the addition compound

$$\underset{CH_3CH_2O}{\overset{CH_3}{\diagdown}}\underset{\diagup}{\overset{\diagdown}{C}}\underset{CN}{\overset{OH}{\diagup}}, \text{ when treated with HCN?}$$

D. *The aldol condensation*

When acetaldehyde is treated with a cold, very dilute aqueous solution of sodium carbonate, the molecule dimerizes to yield a compound containing both an aldehyde and an alcohol group, called aldol.

$$2CH_3CHO \;\underset{}{\overset{\text{very dilute } OH^{\ominus}}{\rightleftharpoons}}\; CH_3{-}\underset{\underset{OH}{|}}{CH}{-}CH_2{-}CHO$$

aldol (systematic name, 3-hydroxybutanal)

This type of reaction is common to most aldehydes and, to a lesser extent, ketones. It has a wide application in synthetic organic chemistry.

The reaction is found to proceed only where the reacting molecule has at least one hydrogen atom α to the carbonyl group. Such hydrogen atoms, as we have seen for acetone (page 16), are 'active'; they are very slightly acidic and can be removed as protons by a suitable base to yield a resonance stabilized carbanion. For acetaldehyde, this sequence is

$$H-\overset{\ominus}{\ddot{O}}\overset{..}{:} \quad H \quad CH_2-CHO \rightleftharpoons H_2O + \left[\overset{\ominus}{C}H_2-C=O \leftrightarrow CH_2=C-\overset{\ominus}{O} \right]$$

The carbanion so produced will be a very effective nucleophile, and as the equilibrium for its formation will lie significantly to the left, it can attack the carbonyl group of a neutral acetaldehyde molecule to form a new stable carbon–carbon bond:

$$\begin{array}{c} CH_3 \\ \diagdown \overset{\delta+}{C} \overset{\delta-}{=} O \\ H \diagup \quad \overset{\ominus}{} \\ :CH_2-CHO \end{array} \quad \xrightarrow{\text{fast}} \quad \begin{array}{c} CH_3 \\ \diagdown C-O^{\ominus} \\ H \diagup \diagdown CH_2-CHO \end{array}$$

The oxy-anion so produced will now easily accept a proton from a water molecule to form aldol:

$$\begin{array}{c} CH_3 \quad O^{\ominus} \\ \diagdown \diagup \\ C \\ \diagup \diagdown \\ H \quad CH_2CHO \end{array} + H_2O \rightarrow \begin{array}{c} CH_3 \quad OH \\ \diagdown \diagup \\ C \\ \diagup \diagdown \\ H \quad CH_2CHO \end{array} + OH^{\ominus}$$

Hydroxyl ion is regenerated, and hence exerts a catalytic effect in the reaction.

Barium hydroxide is the usual catalyst for the dimerization of acetone. Here the product, diacetone alcohol, is formed less rapidly, and in order to drive the equilibrium in its favour, it is normally continually removed from the sphere of reaction as it is formed.

$$2CH_3.CO.CH_3 \underset{\xleftarrow{\hspace{1cm}}}{\overset{\text{solid Ba(OH)}_2}{\xrightarrow{\hspace{1cm}}}} \overset{\overset{\displaystyle OH}{|}}{CH_3.CO.CH_2-C(CH_3)_2}$$
$$\text{diacetone alcohol}$$

Questions

1 In the formation of diacetone alcohol from acetone
  (a) write down the formula of the attacking nucleophile.
  (b) suggest why the reaction is slower than the similar reaction for acetaldehyde.

2 Name a possible by-product in the iodoform reaction (I_2/Na_2CO_3) of acetaldehyde.

3 Suggest a possible synthesis for $HOCH_2$—CH_2—CHO. (Hint: formaldehyde has no α-hydrogen atoms.)

SUMMARY

Aldehydes and ketones undergo addition reactions by nucleophilic attack on carbon due to the inherent polarisation of the isolated carbonyl group. Other carbonyl compounds do not readily undergo addition reactions because resonance of the carbonyl group with neighbouring lone electron pairs diminishes the group's polarity.

3.3 Of nitriles

STRUCTURE

Nitriles are organic cyanides, R—CN. They have significant dipole moments (3–4 D) indicating a marked degree of polarization. In aromatic nuclei the —CN group is electron withdrawing, deactivating and meta directing (see table page 40). The cyanide group, which is most frequently represented by a triple nitrogen–carbon bond, therefore lies between canonical structures

$$-C{\equiv}N: \ \leftrightarrow \ -\overset{\oplus}{C}{=}\overset{..}{N}:^{\ominus} \qquad \text{i.e.} \qquad \overset{\delta+}{-C}{\equiv}\overset{\delta-}{N}$$

Hydrolysis to Carboxylic Acids

When nitriles are refluxed with aqueous mineral acids or alkalis of medium strength, they yield as the final product the corresponding carboxylic acid or its salt. The reaction proceeds through the amide, but this intermediate is usually difficult to isolate.

$$R{-}C{\equiv}N + H_2O \xrightarrow{\ H^{\oplus} \text{ or } OH^{\ominus}\ } R{-}\underset{\underset{O}{\|}}{C}{-}NH_2$$

amide

then

$$R{-}\underset{\underset{O}{\|}}{C}{-}NH_2 + H_2O \xrightarrow{\ H^{\oplus} \text{ or } OH^{\ominus}\ } R{-}\underset{\underset{O}{\|}}{C}{-}OH + NH_3$$

acid

6

for example:

$$CH_3CH_2\!-\!CN + 2H_2O \;\rightarrow\; CH_3CH_2COOH + NH_3$$

The route to the amide will depend upon the catalytic species used. The dipolar nature of the cyanide group will render it prone either to the addition of a proton on nitrogen, or to the addition of hydroxyl ion on carbon. Thus for ethyl cyanide (propionitrile) the first step will be

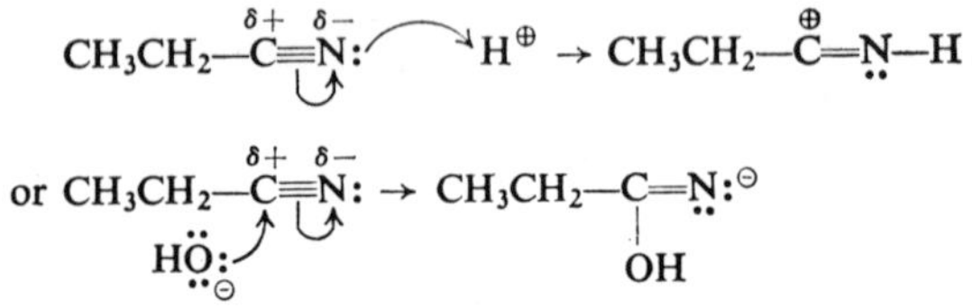

Subsequent reaction of these ions with the solvent, followed by $H^\oplus$ or $OH^\ominus$ loss, yields the same imino alcohol:

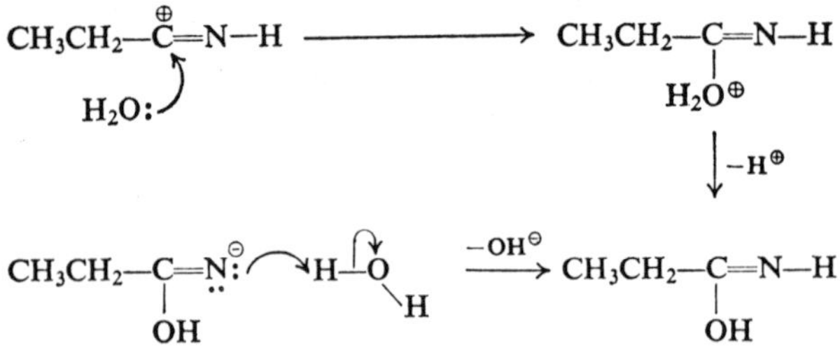

the imino alcohol

Proton transfer in the imino alcohol yields the amide:

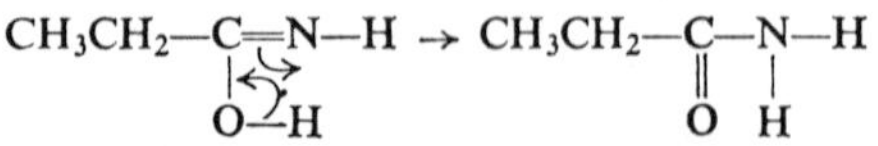

propionamide

By this stage, the carbon–nitrogen bond has been effectively hydrated. The subsequent reaction is the acid or base catalyzed hydrolysis of an amide; these processes are very similar in type to the reversal of acid catalyzed esterification (section 2·4) and ester saponification (question (1), page 26) respectively.

We have mentioned, (page 67), that amides, like esters, are less susceptible to nucleophilic attack than aldehydes and ketones due to the overruling resonance effect:

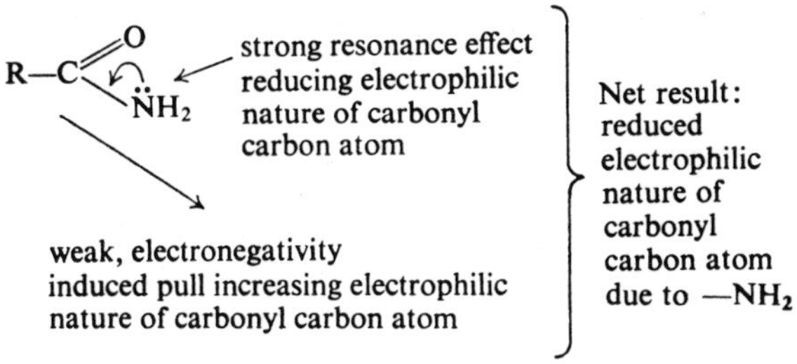

For nucleophilic attack on the carbonyl carbon to take place we therefore require either, the initial addition of an entity which will take the nitrogen lone electron pair out of circulation, or a very powerful nucleophile. This is in fact what takes place in the acid and alkaline hydrolysis reactions.

Acid hydrolysis:

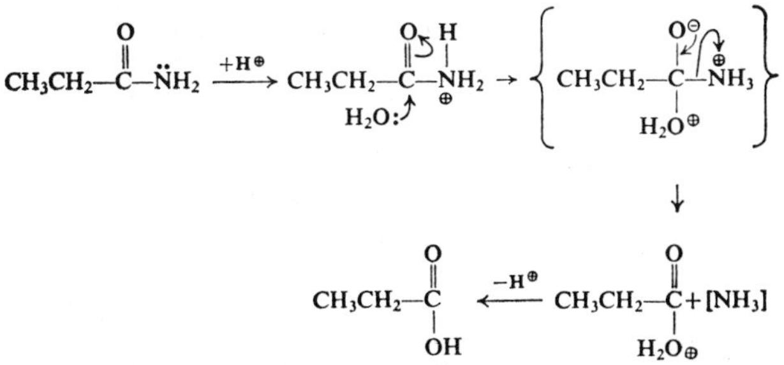

Here, protonation of the nitrogen atom inhibits the resonance effect by localizing the lone electron pair. Nucleophilic attack by water on the carbonyl carbon atom (now significantly positive) then takes place to form the transient addition intermediate. This rapidly decomposes to yield ammonia, which is converted to the ammonium salt of the acid used, and propionic acid (after deprotonation). Using aqueous hydrochloric acid, the overall equation from the amide is

$$CH_3CH_2CONH_2 + HCl + H_2O \rightarrow CH_3CH_2COOH + NH_4Cl$$

Alkaline hydrolysis:

$$CH_3CH_2{-}\overset{\overset{\displaystyle O}{\|}}{C}{-}NH_2 \;\; \underset{\underset{\ominus}{HO:}}{\longrightarrow} \;\; \left\{ CH_3CH_2{-}\underset{\underset{\displaystyle HO}{|}}{\overset{\overset{\displaystyle O^{\ominus}}{|}}{C}}{-}NH_2 \right\} \;\; \longrightarrow \;\; CH_3CH_2{-}\underset{\underset{\displaystyle HO}{|}}{\overset{\overset{\displaystyle O}{\|}}{C}}{+}N\overset{\ominus}{H_2}$$

$$\downarrow$$

$$CH_3CH_2{-}COO^{\ominus}+NH_3\uparrow$$

In this case $OH^{\ominus}$ is the powerful nucleophile which slowly reacts with the amide to form the addition intermediate. This species quickly breaks apart to form propionic acid and the $NH_2{}^{\ominus}$ ion. In the alkaline medium, the carboxylic acid will naturally exist in salt form, and ammonia will be liberated by the rapid acquisition of a proton by $NH_2{}^{\ominus}$ either from the carboxylic acid or from a water molecule. An overall equation would be

$$CH_3CH_2CONH_2+Na^{\oplus}OH^{\ominus} \;\rightarrow\; CH_3CH_2COO^{\ominus}Na^{\oplus}+NH_3\uparrow$$

Questions

1 Write down the formulae of compounds produced by the reaction of $CH_3{-}CH{=}CH{-}CN$ with (a) Pt/H_2, (b) $LiAlH_4$.
2 Suggest a synthesis of glycollic acid, $HO{-}CH_2{-}COOH$ from formaldehyde.
3 Outline a sequence for the conversion of phenyl cyanide, $C_6H_5{-}CN$, to meta-nitrobenzoic acid.

NOTE ON THE REACTIONS OF CARBONYL COMPOUNDS

From an assessment of the reactions of carbonyl compounds encountered in various parts of the foregoing text we can see an essential unity. All are attacked by the same general types of reagent under different conditions. That the carboxylic acids and their derivatives (esters, amides, acid chlorides, etc.) do not form simple stable addition compounds, as do the aldehydes and ketones, is the consequence of the essential reversibility of the addition reaction. The addition intermediate in the case of acid derivatives, which has been shown in brackets {   } in the text, reverts to a new product by a net substitution reaction, instead of to the original substances from which it arose.

4 Elimination Reactions

4.1 The internal dehydration of ethanol

Under suitable conditions, ethanol can be dehydrated with concentrated sulphuric acid to ethylene:

$$CH_3\!-\!CH_2\!-\!OH \;-\; H_2O \xrightarrow[H_2SO_4]{conc.} CH_2\!\!=\!\!CH_2$$

The single carbon–carbon bond is transformed into an olefinic double bond containing two extra electrons.

The first stage of the reaction will be protonation of the nucleophilic alcohol:

$$CH_3\!-\!CH_2\!-\!OH + H_2SO_4 \;\rightleftharpoons\; CH_3\!-\!CH_2\!-\!\overset{\oplus}{O}H_2 + HSO_4{}^{\ominus}$$

compare

$$H\!-\!OH + H_2SO_4 \;\rightleftharpoons\; \overset{\oplus}{H}OH_2 + HSO_4{}^{\ominus}$$

This process is entirely analogous to the protonation of water by sulphuric acid and the equilibrium lies well to the right. It is now possible for the much weakened carbon–oxygen bond to break, the protonated alcohol dissociating thus:

$$CH_3\!-\!CH_2\!\!-\!\!\overset{\oplus}{O}H_2 \;\rightarrow\; CH_3\!-\!\overset{\oplus}{C}H_2 + H_2O$$

A highly reactive positively charged carbonium ion is produced. The positive charge is centred on the carbon atom shown, this atom possessing six outer shell electrons. The lifetime of this cation will be very short, and it rapidly stabilizes itself by dropping off a proton from the adjacent carbon atom.

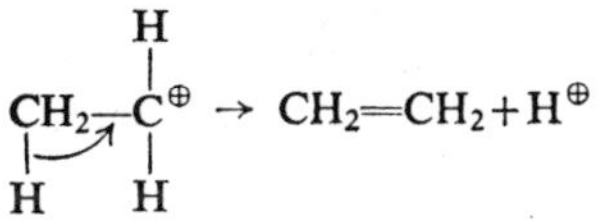

$$\underset{H}{\overset{H}{CH_2\!-\!\overset{|}{\underset{|}{C}}{}^{\oplus}}} \;\rightarrow\; CH_2\!\!=\!\!CH_2 + H^{\oplus}$$

This last stage of deprotonation will be facilitated if there is a suitable base (proton acceptor) in the reaction mixture. Bisulphate ion, the conjugate base of sulphuric acid, is present in high concentration, as the reaction conditions to produce a good yield of ethylene are excess sulphuric acid at 170°C. The last two stages of the reaction can therefore be represented by the following concerted electron pair shifts:

$$CH_2{-}CH_2{-}\overset{\oplus}{O}H_2 \rightarrow \underset{O}{\overset{O}{>}}S\underset{OH}{\overset{OH}{<}} + CH_2{=}CH_2 + H_2O$$

H (protonated
ethanol)

$\underset{O}{\overset{O}{>}}S\underset{OH}{\overset{\cdot\cdot\ominus}{<}}$ Nucleophilic attack
on hydrogen

(bisulphate ion)

In this formulation the ethyl carbonium ion is not shown as an intermediate, as deprotonation is synchronous with the elimination of water from the protonated ethanol molecule.

It should be emphasized that all these steps in the reaction are reversible, and in fact, ethylene can be converted back to ethanol in reasonable yield by bubbling the gas into concentrated sulphuric acid at 80°C and then heating the solution with excess water at 100°C*. In the sequence above, the ethylene (b.p. $-102°C$) distills out from the reaction mixture as it is formed, thus displacing the equilibria in favour of the olefin.

There are several other ways in which the protonated ethanol entity can react with the species present in the ethanol–sulphuric acid reaction mixture. These are all substitution reactions, and they are favoured over the 1,2 elimination reaction described above (adjacent carbon atoms lose an atom or group of atoms) by lower reaction temperatures, for example:

1. At low temperatures (ca. 0°C) with excess sulphuric acid ethyl hydrogen sulphate, $C_2H_5OSO_2OH$, is formed. If the reaction mixture containing this substance is heated to about 100°C under reduced pressure, diethyl sulphate, $C_2H_5OSO_2OC_2H_5$, distills out. In these reactions, sulphuric acid is consumed.

2. At about 140°C with excess ethanol, diethyl ether, $C_2H_5OC_2H_5$, is the main product. Here, as in the reaction to produce ethylene, no sulphuric acid is used up, the process being termed the 'continuous etherification process'.

The reaction of ethanol with concentrated sulphuric acid clearly illustrates the vital role of practical conditions in organic reactions. In order to obtain the maximum yield of the desired product, these conditions have frequently to be very carefully selected.

* Ethanol is now made on an industrial scale by reaction of ethylene and hot aqueous sulphuric acid under pressure.

Questions

1 Formulate electron pair shifts to account for the formation of (a) ethyl hydrogen sulphate, (b) diethyl sulphate, and (c) diethyl ether in an ethanol–sulphuric acid reaction mixture. Explain briefly.
2 Ethylene is formed from ethanol by an internal or intramolecular dehydration process. Indicate the type of dehydration undergone by ethanol when diethyl ether is produced.
3 Show clearly the reaction sequence for the hydration of ethylene with aqueous acid. Name the three steps in the reaction.

4.2 The dehydration of aldol

We have seen (section 3·2D) that aldol, $CH_3CH(OH)CH_2CHO$, is formed by the dimerization of two acetaldehyde molecules at low temperatures in the presence of a trace of hydroxyl ion (from, for example, 1 per cent aqueous sodium carbonate). If the reaction solution is subsequently heated, a further reaction occurs resulting in the formation of crotonaldehyde by the elimination of water from aldol:

$$CH_3\text{—}\underset{\underset{OH}{|}}{CH}\text{—}CH_2\text{—}CHO \xrightarrow{-H_2O} \underset{\text{crotonaldehyde}}{CH_3\text{—}CH\text{=}CH\text{—}CHO}$$

This is essentially the dehydration of an alcohol, but in contrast to the reaction described in the previous section, this dehydration can be effected in the presence of bases as well as acids. Here we encounter once again the activating effect of a lone carbonyl group on the next-door (α) hydrogen atoms. These hydrogen atoms are slightly acidic and can therefore be readily abstracted in the presence of a base to give a resonance stabilized carbanion. Thus, for aldol we have:

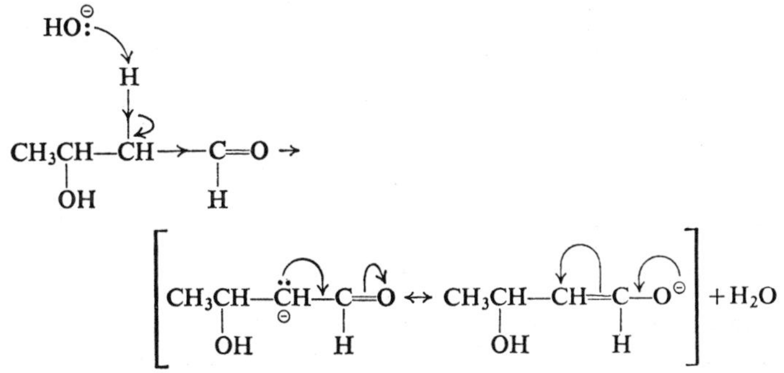

Although this carbanion has some stability, it gains permanent stability by dropping off $OH^{\ominus}$ to yield crotonaldehyde:

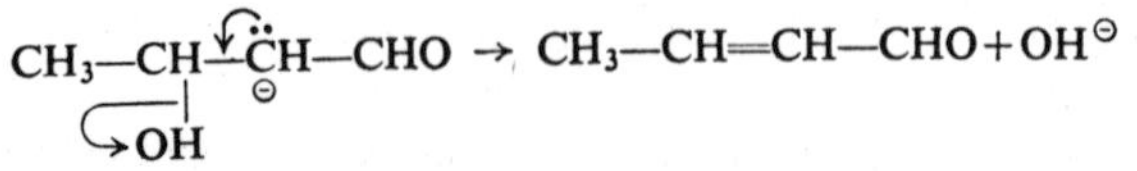

$$CH_3-CH-CH-CHO \rightarrow CH_3-CH=CH-CHO+OH^{\ominus}$$
$$\overset{|}{\underset{OH}{}}$$

Here the overall dehydration reaction has been pictured as a two stage process. It is not certain whether the carbanion has any significant existence, and the reaction can be equally well represented by a concerted electron movement:

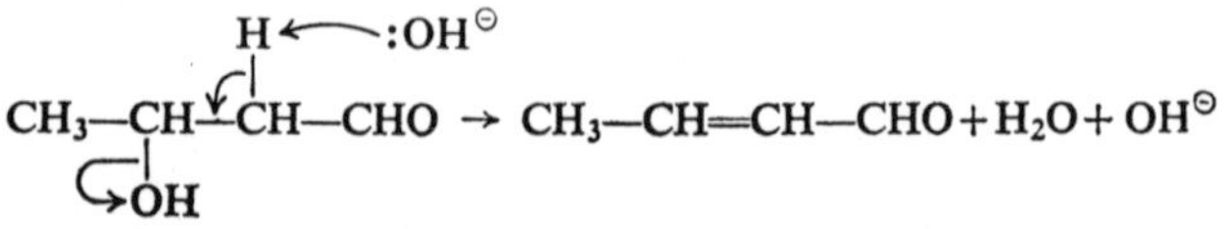

$$H\longleftarrow :OH^{\ominus}$$
$$CH_3-CH-CH-CHO \rightarrow CH_3-CH=CH-CHO+H_2O+OH^{\ominus}$$
$$\overset{|}{\underset{OH}{}}$$

It is possible that some of the driving force for this reaction may be due to the resonance stabilization of the product

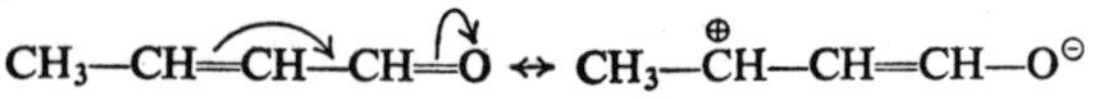

$$CH_3-CH=CH-CH=O \leftrightarrow CH_3-\overset{\oplus}{C}H-CH=CH-O^{\ominus}$$

This representation implies that there is some overlapping of the two π electron systems of crotonaldehyde:

$$^4CH_3-{}^3CH-{}^2CH-{}^1CH-O$$

The measured C_1-C_2 bond length of $1 \cdot 46$ Å confirms the intermediate nature of this bond indicated by the resonance structures (compare $C-C$ $1 \cdot 54$ Å; $C=C$ $1 \cdot 34$ Å).

By analogy with the dehydration of ethanol, the acid catalysed elimination of water from aldol (by warm dilute acid) can be pictured:

$$CH_3-CH-CH_2-CHO+H^{\oplus} \rightarrow CH_3-CH-CH_2-CHO$$
$$\overset{|}{\underset{OH}{}} \qquad\qquad\qquad \overset{|}{\underset{\oplus OH_2}{}}$$

then

$$H\longleftarrow :OH_2$$
$$CH_3-CH-CH-CHO \rightarrow CH_3-CH=CH-CHO+H_2O+H_3O^{\oplus}$$
$$\overset{|}{\underset{\oplus OH_2}{}}$$

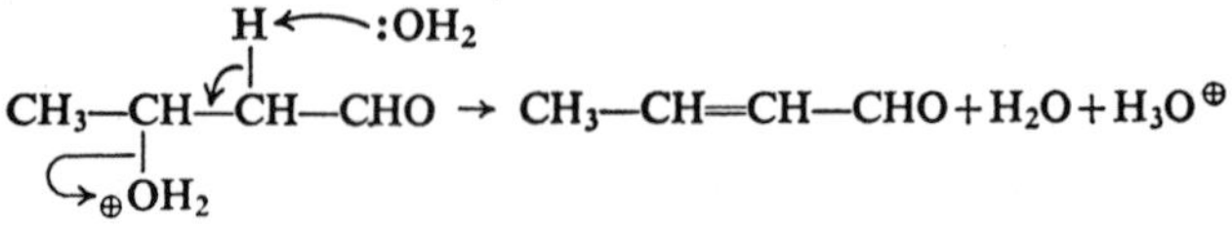

that is, protonation, followed by nucleophilic attack on hydrogen. The relative ease with which the dehydration occurs, compared with ethanol, is chiefly a result of the reactivity of the acidic α-hydrogen atom, especially in the protonated entity, towards even fairly weak nucleophiles.

Questions

1 Write down the formula of the compound produced when diacetone alcohol, $(CH_3)_2C(OH)CH_2COCH_3$, is heated with a trace of dilute sulphuric acid.

2 The central carbon–carbon bond length of butadiene, $CH_2{=}CH{-}CH{=}CH_2$, is 1·46 Å. Explain.

3 The overall equation for the 'condensation' of aldehydes and ketones with derivatives of hydrazine is

$$RR'C{=}O + NH_2{-}NHR'' \rightarrow RR'C{=}N{-}NHR'' + H_2O$$

These reactions occur readily at room temperature at an intermediate pH and they proceed by a similar pathway to the total aldol reaction (that is, addition followed by elimination—see page 3). How does nitrogen initially become attached to the carbonyl carbon atom? Write down the formula of the product obtained by treating acetone with phenyl hydrazine.

4.3 Dehydrobromination

When alkyl halides are treated with a concentrated solution of potassium hydroxide in ethanol as solvent, some dehydrohalogenation takes place to form an olefin. The yield of olefin depends very greatly upon the structure of the alkyl halide and the actual reaction conditions. As an example, we will consider the reaction of isopropyl bromide which yields about 80 per cent propylene in the reaction:

$$CH_3.CHBr.CH_3 + K^{\oplus}OH^{\ominus} \xrightarrow[\text{alcohol}]{\text{heat in}} CH_3.CH{=}CH_2 + K^{\oplus}Br^{\ominus} + H_2O$$

At some stage in the reaction, bromine is liberated as its anion, and as usual, the reaction can be depicted in terms of a two stage process (I), or by a single concerted mechanism (II).

(I)

$$H{-}CH_2{-}\overset{\overset{\displaystyle CH_3}{|}}{CH}{-}Br \xrightarrow[\text{slow}]{-Br^{\ominus}} H{-}CH_2{-}\overset{\overset{\displaystyle CH_3}{|}}{\overset{\oplus}{CH}} \xrightarrow[\text{fast}]{-H^{\oplus}} \overset{\overset{\displaystyle CH_3}{|}}{CH_2{=}CH}$$

carbonium ion $(H^{\oplus}+OH^{\ominus} \rightarrow H_2O)$

(II)

$$HO{\overset{\ominus}{\colon}} \rightarrow H{-}CH_2{-}\overset{\overset{\displaystyle CH_3}{|}}{CH}{-}Br \xrightarrow{\text{slow}} H_2O + \overset{\overset{\displaystyle CH_3}{|}}{CH_2{=}CH} + Br^{\ominus}$$

Process (I) would give a kinetic expression independent of the reagent:

Rate α [CH$_3$.CHBr.CH$_3$]   (unimolecular)

whereas process (II) would give the expression

Rate α [CH$_3$.CHBr.CH$_3$] [OH$^\ominus$] (bimolecular)

In practice, for isopropyl bromide, propylene will be being formed by both routes concurrently in proportions which are dependent upon the precise reaction conditions, though in concentrated alkali, process (II) is the more important.

COMPETITION OF ELIMINATION WITH SUBSTITUTION REACTIONS

The elimination reaction here described is however in direct competition with possible substitution reactions. Alkyl halides, having a potentially good leaving group in halogen anion, are prone to nucleophilic displacement reactions (see section 2.2). In alcoholic potash both hydroxyl and ethoxide ion will be present in the solution due to the equilibrium

$$C_2H_5OH + K^\oplus OH^\ominus \rightleftharpoons C_2H_5O^\ominus \; K^\oplus + H_2O$$

Reaction of ethyl bromide with hot alcoholic potassium hydroxide in fact yields only 2 per cent of the elimination product, ethylene. By far the most predominant product in this case is diethyl ether produced by the Williamson synthesis (nucleophilic displacement of Br$^\ominus$ by C$_2$H$_5$O$^\ominus$, page 13). This reaction is clearly not a satisfactory method for preparing ethylene. On the other hand, tertiary butyl bromide gives almost 100 per cent of the olefin isobutylene under similar conditions:

$$CH_3 - \underset{\underset{CH_3}{|}}{\overset{\overset{CH_3}{|}}{C}} - Br \xrightarrow[(-HBr)]{\text{hot KOH/alcohol}} CH_3 - \underset{\underset{CH_2}{\|}}{\overset{\overset{CH_3}{|}}{C}}$$

In this case there is almost no tendency for substitution reactions to take place.

A marked difference in the course of the reaction is therefore observed which depends upon the degree of substitution at the carbon atom bearing the bromine atom. Increasing substitution at this carbon atom favours the elimination reaction, as the table shows.

Reaction of simple alkyl bromides with hot concentrated alcoholic KOH:

Compound	Formula	Type	Approx. per cent yield of olefin	Reaction Course
ethyl bromide	CH_3—CH_2Br	primary	2	substitution
isopropyl bromide	$CH_3.CHBr.CH_3$	secondary	80	chiefly elimination
t. butyl bromide	$(CH_3)_3CBr$	tertiary	100	elimination

This effect is chiefly due to the influence of steric factors. Methyl and other alkyl groups are quite large, and tertiary bromides will undergo a significant release of spatial strain when the bulky bromine atom is 'pushed out' in the reaction. In addition, the three bulky alkyl groups will prevent the close approach of any possible incoming substituent ($OH^\ominus$ or $C_2H_5O^\ominus$) to the tertiary carbon atom. The elimination reaction, which reduces spatial compression, is therefore strongly favoured over possible substitution reactions, which would reintroduce steric strain, in the case of tertiary halides. Primary halides favour substitution reactions and secondary halides are generally intermediate in behaviour.

STERIC COURSE OF REACTION

Using stereoisomeric substrates, it has been possible to investigate the steric course of the elimination reaction. The more important bimolecular reaction path has been found to be surprisingly selective. For example, the base induced elimination of hydrogen bromide from bromofumaric acid to yield acetylene dicarboxylic acid proceeds much more rapidly than the corresponding elimination from its geometrical isomer, bromomaleic acid.

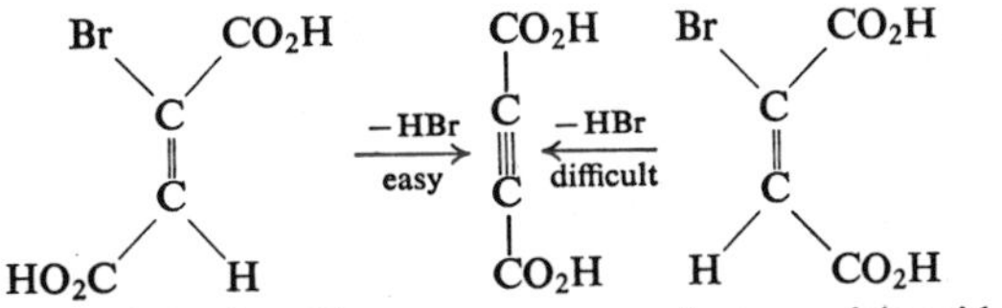

bromofumaric acid   bromomaleic acid

This suggests that a *trans* elimination process is the more favoured reaction; the trans isomer has the hydrogen and bromine atoms on opposite sides of the molecule (no rotation about $>C{=}C<$) and a rapid elimination can result:

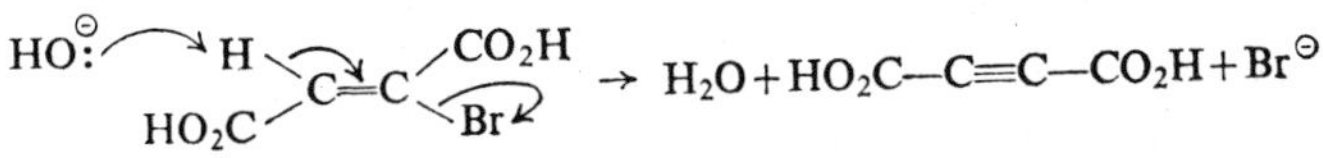

This is confirmed by the fact that single dehydrobromination of dl. and meso 2,3-dibromosuccinic acids (see page 54) yields respectively bromo-fumaric and bromomaleic acids:

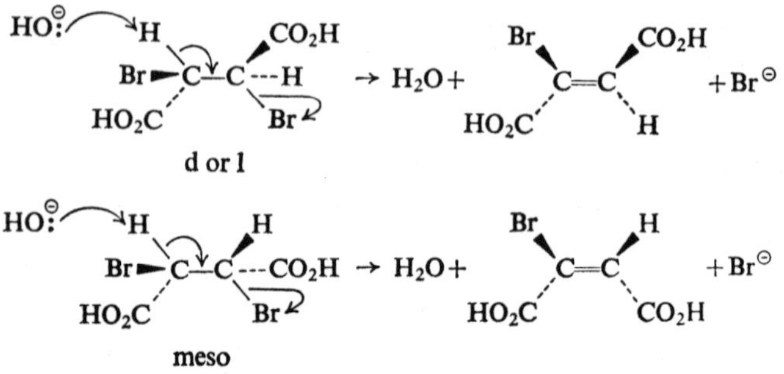

d or l

meso

The speeds of these reactions are comparable because the atoms to be eliminated can arrange themselves for reaction in a *trans* coplanar dis-position by free rotation about the *single* central carbon–carbon bond.

Questions

1 Give the formula of an alternative entity to $OH^{\ominus}$ in the reactions described in this section.

2 Give the formulae of possible by-products in the reaction of isopropyl bromide with hot concentrated alcoholic KOH, stating how they are likely to arise.

3 In the reaction of dl.2,3-dibromosuccinic acid with base shown above, only one enantiomorph is shown. Give clearly the reaction path for its mirror image, drawing the H—C and C—Br bonds to be ruptured in the plane of the paper. (Ball and stick models will be found useful in studying this.)

4 Indicate how ethylene would arise by warming the quaternary ammonium compound $CH_3CH_2\overset{\oplus}{—}N(CH_3)_3I^{\ominus}$, (obtained from ethylamine with excess methyl iodide), with aqueous alkali.

4.4 The debromination of 1,2 dibromides with zinc

On refluxing a 1,2 dibromide with zinc dust in alcohol as solvent, both bromine atoms are removed to yield an olefin; for example:

$$CH_3—CHBr—CHBr—CH_3 + Zn \rightarrow CH_3—CH=CH—CH_3 + ZnBr_2$$

2,3-dibromobutane   2-butene

The reaction is therefore the reversal of the addition of bromine to an olefin. The combined reactions are often used to purify olefinic compounds, the intermediate dibromides (section 3.1A) being readily isolated crystalline solids or high boiling liquids.

The divalent metal zinc possesses a pair of electrons in its outermost shell which it is able to donate to a suitable receptor atom. Bromide anions are formed in the reaction solution, and the reaction can be pictured by a concerted process:

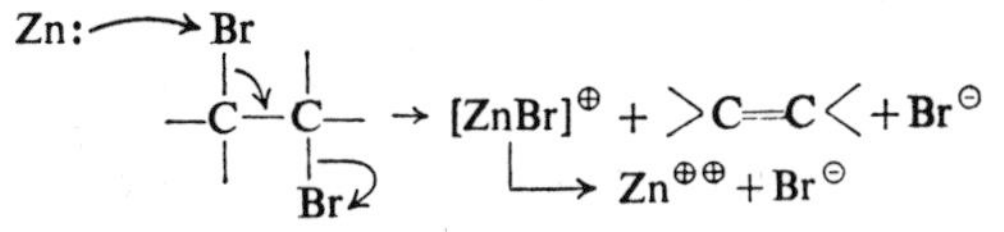

Zinc here is effectively acting as a nucleophile and the first bromine atom comes off as $Br^{\oplus}$. Again, the elimination has been shown to occur by a *trans*-mechanism.

Questions

1 How would you separate $(CH_3)_2C{=}CHCH_3$ (b.p. 38°C) from its saturated analogue $(CH_3)_2CH.CH_2.CH_3$ (b.p. 28°C)?

2 Name the precise product formed when meso 2,3 dibromobutane is treated with Zn/C_2H_5OH.

SUMMARY

Elimination reactions to form olefins can proceed either by a one stage concerted process involving both reagent and substrate (bimolecular), or by a two stage process in which the reagent plays no part in the rate controlling step (unimolecular). The concerted process involves attack by a nucleophile on one of the atoms (often hydrogen) to be eliminated.

Most eliminations proceed via a trans mechanism.

5 Rearrangement Reactions

5.1 The Hofmann degradation of amides

Careful treatment of an acid amide, $RCONH_2$ ($R = $ alkyl), with bromine in aqueous alkali results in the formation of an amine, RNH_2. At some stage during the reaction a new R—N bond is formed and one carbon atom is 'lost' as carbon dioxide which is converted directly to carbonate ion in the alkaline reaction medium. The overall equation for the conversion of acetamide to methylamine ($R = CH_3$) can be given as

$$CH_3CONH_2 + Br_2 + 4OH^\ominus \rightarrow CH_3NH_2 + CO_3^{\ominus\ominus} + 2Br^\ominus + 2H_2O$$

sodium or potassium hydroxide being the alkalis usually employed. Yields are often excellent.

Under suitable conditions, it is possible to isolate three different intermediate species in the reaction. These are

RCONHBr	the N-bromoamide
RCO$\overset{\ominus}{N}$Br	in salts of the N-bromoamide
R—N=C=O	the isocyanate

These species can all undergo further reaction with the reagent to yield the amine as the final product. The reaction path must therefore be formulated so as to proceed through these intermediate species.

The first step consists of the formation of the bromoamide by substitution of bromine for one of the hydrogen atoms which is attached to nitrogen and is in a position to be activated by the adjacent carbonyl group. This reaction occurs in the cold.

$$CH_3CONH_2 + Br_2 + OH^\ominus \rightarrow CH_3CONHBr + Br^\ominus + H_2O$$

N-bromoamides are stable in the absence of excess alkali, but in the Hofmann reaction a significant excess is employed. The second step in the reaction is therefore deprotonation of the bromoamide in the presence of alkali to form its anion.

$$CH_3CONHBr + OH^\ominus \rightarrow CH_3CO\overset{\ominus}{N}Br + H_2O$$

This relatively unstable anion now undergoes reaction involving the rearrangement step at about 70°C to form the new R—N bond present in the intermediate isocyanate.

This is here pictured as elimination of bromide anion to form a highly unstable neutral entity $CH_3CO\overset{..}{N}$, which only has a sextet of valency electrons round the nitrogen atom. This entity is therefore highly electron deficient on nitrogen, and it gains some stability by the migration of the methyl group *with its pair of bonding electrons* to form a formal dipolar ion which is more truly represented by the isocyanate structure. All valency electron octets are now complete. It is uncertain whether the electron deficient nitrogen species has an independent existence, for it could, in aqueous solution, react with water to yield compounds of the type

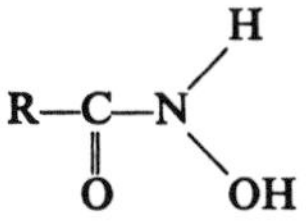

$$R-\underset{\underset{O}{\|}}{C}-N\overset{H}{\underset{OH}{\diagdown}} \quad \text{(valency octets preserved)}$$

which are known to be stable. No such compounds have ever been isolated from Hofmann reaction mixtures. For this reason a simultaneous departure of bromide anion and shift of the migrating group is probably a more accurate picture for this rate-controlling stage of the reaction:

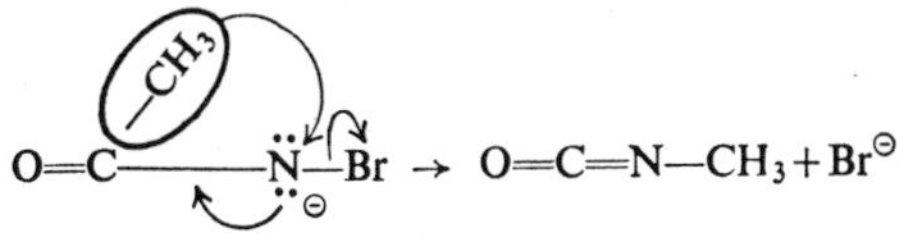

$$O{=}C{-}\overset{..}{N}{-}Br \rightarrow O{=}C{=}N{-}CH_3 + Br^{\ominus}$$

The three stage process shown previously does however serve to emphasize the mechanistic basis for the rearrangement.

The isocyanate is quickly converted to the amine at 70°C by base catalysed hydration of the carbon–nitrogen double bond, followed by spontaneous decarboxylation (loss of CO_2):

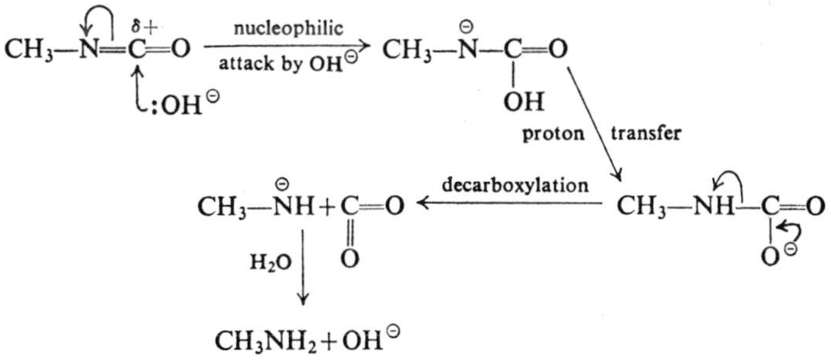

$$CH_3-\overset{\delta+}{N}{=}C{=}O \xrightarrow[\text{attack by OH}^\ominus]{\text{nucleophilic}} CH_3-\underset{OH}{N}-C{=}O$$

$$:OH^\ominus \qquad\qquad \text{proton transfer}$$

$$CH_3-\overset{\ominus}{N}H + C{=}O \xleftarrow{\text{decarboxylation}} CH_3-NH-C{=}O$$

$$H_2O \mid \quad O$$

$$CH_3NH_2 + OH^\ominus$$

$$(CO_2 + 2OH^\ominus \rightarrow CO_3^{\ominus\ominus} + H_2O)$$

88

It has been established that Hofmann reactions are accelerated if the migrating group R has an increased electron releasing capacity. A strongly electron donating migratory group not only eases the departure of $Br^\ominus$ from the bromoamide anion, but is also able to satisfy the electron deficiency of the residual nitrogen atom more effectively. Thus, the rate of amine formation from p-hydroxybenzamide

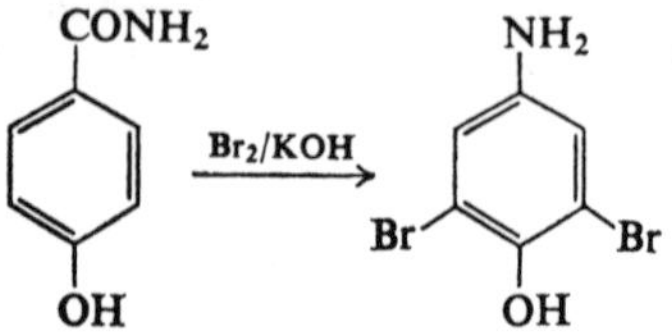

is more rapid than that for benzamide itself due to the activating effect of the phenolic-OH group. Bromination into the aromatic ring has also occurred in this example, again due to activation by the phenolic —OH).

Questions

1 The N-substituted amide $CH_3CONHCH_3$ does not yield the secondary amine CH_3NHCH_3 upon treatment with Br_2/KOH. This supports the mechanism outlined above. Explain, and give the formula of the compound which will form.
2 Say what you can about the relative rates of amine formation from benzamide and p-nitrobenzamide when treated under identical conditions with $Br_2/NaOH$.

5.2 The rearrangement of N-chloroacetanilide

N-chloroacetanilide is prepared from aniline by the following reaction sequence:

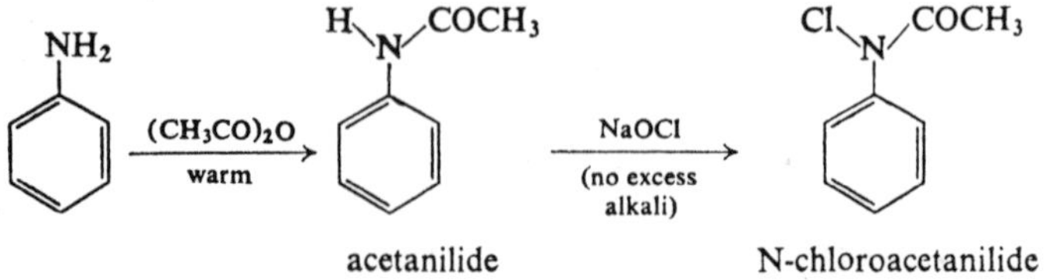

acetanilide   N-chloroacetanilide

The first stage is acetylation of the amine, in which the lone pair of electrons on nitrogen act as a nucleophile towards the carbonyl carbon atom. The second step is analogous to the first stage of the Hofmann degradation of amides (section 5.1).

If N-chloroacetanilide is treated with aqueous hydrochloric acid, the chlorine atom migrates to form a mixture of ortho and para chloroacetanilides:

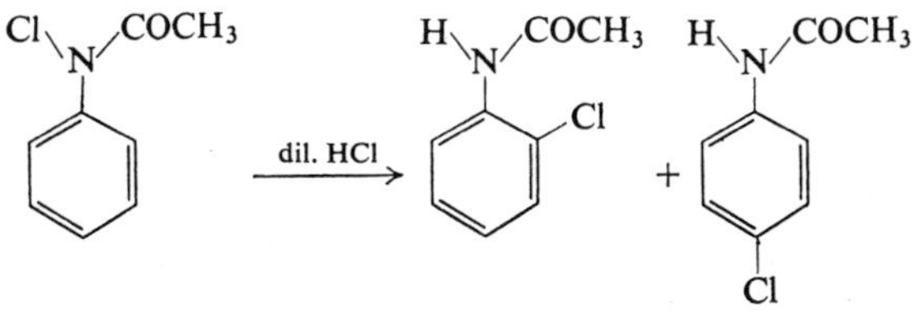

This, at first sight remarkable reaction, has been extensively investigated and the mechanism is now certain. The evidence is as follows:

(a) The reaction is specific for hydrochloric acid, and if air is rapidly bubbled through the solution whilst the reaction is taking place, chlorine is evolved to leave a residue of acetanilide.

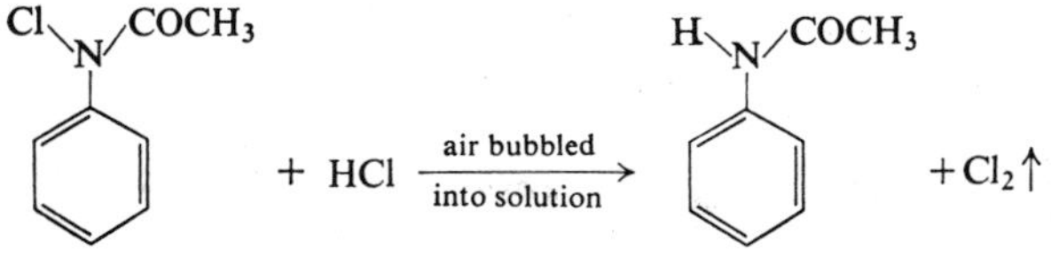

This suggests that chlorine and acetanilide molecules are involved as intermediate species in the reaction.

(b) The ortho/para ratio of the ring chlorinated acetanilides produced is identical to the ratio obtained when acetanilide (—NHCOCH$_3$ is activating and o/p directing — see table page 40) is treated directly with chlorine in the same solvent.

(c) When the hydrochloric acid used in the reaction is labelled with radioactive chlorine, the ring chlorinated products contain a significant quantity of labelled chlorine in the aromatic ring. This indicates that some chlorine is transferred from the HCl into the organic products.

7

On the basis of this non-kinetic evidence the reaction can now be formulated as proceeding through acetanilide thus:

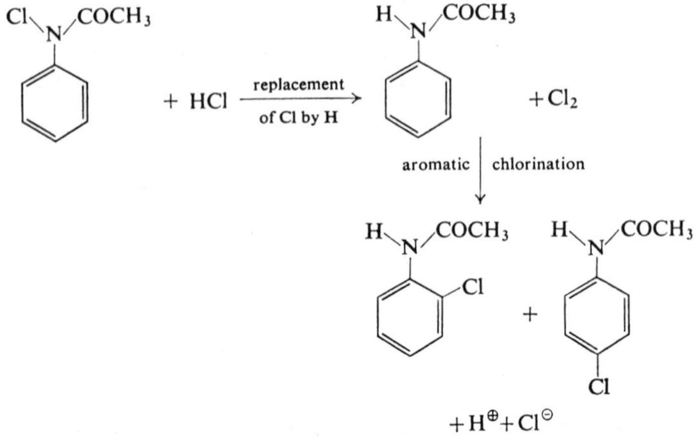

Here, the migrating group, the chlorine atom, becomes completely detached from the remainder of the molecule for a significant period of time during the reaction and combines with $Cl^{\ominus}$ from the acid to form chlorine molecules. The chlorine atom therefore must come off the nitrogen atom as $Cl^{\oplus}$, which suggests prior protonation on nitrogen. The succeeding reaction is a normal electrophilic aromatic chlorination. This stops at the monochlorinated stage because only one molecule of chlorine is produced for every molecule of N-chloroacetanilide which is converted. It can be seen therefore that there is no guarantee that the chlorine atom which returns to a given molecule is that chlorine which has departed from it, and this explains the observed incorporation of labelled chlorine from labelled hydrochloric acid in the final products.

Rate of reaction studies give the kinetic relationship for the reaction

Rate α [N-chloroacetanilide] $[HCl]^2$

that is, a third order reaction. Assuming that the hydrochloric acid is fully ionized and that protonation of the nitrogen atom, suggested above, is in fact the first step of the reaction, this kinetic expression can be modified to

Rate α [N-chloroacetanilide] $[H^{\oplus}]$ $[Cl^{\ominus}]$ (With HCl fully ionized, $[HCl] \equiv [H^{\oplus}]$, and $[H^{\oplus}]=[Cl^{\ominus}]$).

$$\alpha \text{ [C}_6\text{H}_5\text{—}\overset{\overset{\displaystyle H}{|}}{\underset{\underset{\displaystyle Cl}{|}}{N}}{}^{\oplus}\text{—COCH}_3] \text{ [Cl}^{\ominus}]$$

This is therefore in agreement with the proposal that the rate-controlling step in the reaction is nucleophilic attack by $Cl^\ominus$ on the chlorine atom of protonated N-chloroacetanilide:

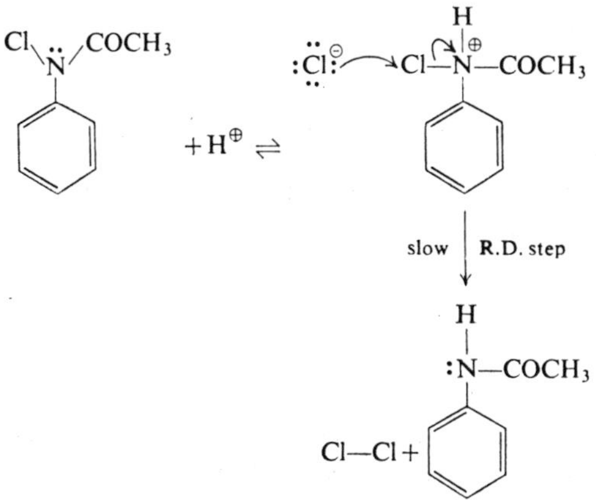

Chlorination of the aromatic ring is then a comparatively rapid process, proceeding via electrophilic attack by, effectively, $Cl^\oplus$ (compare section 2.5C).

Questions

1 Draw clearly the reaction path for the acetylation of aniline by acetyl chloride.

2 Treatment of N-chloroacetanilide with hydriodic acid yields a mixture of ortho and para iodoacetanilides. Explain, and suggest why the para isomer is the predominant product. Now predict the most abundant product formed when N-chloroacetanilide is treated with aqueous HBr.

3 What would happen if the compound

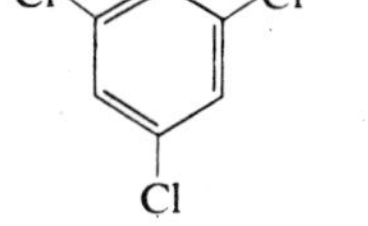

were treated with dilute HCl?

5.3 Allylic rearrangement

Allylic compounds are those which possess a functional group X, other than an unsaturated linkage, on a carbon atom α to an olefinic bond, viz.

$$R-\overset{R}{\underset{X}{\overset{|}{\underset{|}{C}}}}{}^3-\overset{R'}{\underset{}{\overset{|}{C}}}{}^2={}^1C\overset{\nearrow R}{\underset{\searrow R'}{}}$$

This type of structure frequently undergoes acid or base catalyzed functional group migration to yield structures of the type.

$$R-{}^3\overset{R}{\overset{|}{C}}={}^2\overset{R'}{\overset{|}{C}}-{}^1C\overset{\nearrow R'}{\underset{\searrow}{\underset{X\ R}{}}}$$

As an example, we will consider the acid catalyzed rearrangement of 3-methylbuta-l-ene-3-ol ($R = CH_3$, $R' = H$, $X = OH$ above):

$$CH_3-{}^3\overset{CH_3}{\underset{OH}{\overset{|}{\underset{|}{C}}}}-{}^2CH={}^1CH_2 \xrightarrow{H^\oplus} CH_3-\overset{CH_3}{\overset{|}{C}}=CH-CH_2OH$$

This compound can be conveniently synthesized from acetone in two stages:

$$\begin{array}{c}CH_3 \\ \diagdown \\ \diagup \\ CH_3\end{array}C=O \xrightarrow[Na/liq.\ NH_3]{CH\equiv CH} \begin{array}{c}CH_3 \quad OH \\ \diagdown \diagup \\ C \\ \diagup \diagdown \\ CH_3 \quad C\equiv CH\end{array} \xrightarrow[catalyst]{1\ mole\ H_2} \begin{array}{c}CH_3 \quad OH \\ \diagdown \diagup \\ C \\ \diagup \diagdown \\ CH_3 \quad CH=CH_2\end{array}$$

The first stage is one of addition of a molecule of acetylene across the carbonyl double bond. Acetylene is a very weak acid ($pK_a = 26$), and will form the acetylide anion $^\ominus C\equiv CH$ upon treatment with a strong base, like sodamide, in an anhydrous solvent:

$$Na^\oplus NH_2^\ominus + CH\equiv CH \xrightarrow[NH_3]{liq.} Na^\oplus \ ^\ominus C\equiv CH + NH_3$$

Acetylene forms other more stable insoluble metallic salts in aqueous solution, for example, red cuprous acetylide, $Cu-C\equiv C-Cu$, with ammoniacal cuprous chloride, and the cream-coloured silver acetylide, $Ag-C\equiv C-Ag$, with ammoniacal silver nitrate. The precise reasons for the acidity of acetylenic hydrogen atoms are however not at present fully understood. In the addition reaction above, acetylide ion attacks the electron deficient carbonyl carbon atom and the intermediate anion is rapidly protonated from the solvent to yield the acetylenic tertiary alcohol:

$$CH_3 \overset{\delta+}{\underset{CH_3}{C}}=O^{\delta-} \quad \underset{:C\equiv CH}{\ominus} \rightarrow \underset{CH_3}{\overset{CH_3}{C}} \overset{O^{\ominus}}{\underset{C\equiv CH}{}} \xrightarrow{NH_3} \underset{CH_3}{\overset{CH_3}{C}} \overset{OH}{\underset{C\equiv H}{}} + NH_2^{\ominus}$$

The acetylenic triple bond is then partially saturated using hydrogen and Lindlar's catalyst (palladium partially poisoned with a heavy metal salt) to yield the allylic tertiary alcohol.

Rearrangement of the allylic alcohol takes place under the influence of dilute aqueous acid. Protonation of the hydroxyl group followed by loss of water yields an electron deficient carbonium ion:

$$\underset{CH_3}{\overset{CH_3}{C}} \overset{OH}{\underset{CH=CH_2}{}} \xrightarrow{+H^{\oplus}} \underset{CH_3}{\overset{CH_3}{C}} \overset{\oplus OH_2}{\underset{CH=CH_2}{}}$$

$$\downarrow -H_2O$$

$$\underset{CH_3}{\overset{CH_3}{C}} \overset{\oplus}{\underset{CH=CH_2}{}}$$

Elimination of water is naturally favoured in this case as the alcohol is tertiary and any steric congestion is relieved. In addition, the resultant carbonium ion gains some resonance stabilization by delocalization of the π electrons of the olefinic bond:

$$\underset{CH_3}{\overset{CH_3}{C}} \overset{\oplus}{\underset{CH=CH_2}{}} \leftrightarrow \underset{CH_3}{\overset{CH_3}{C^3}} \underset{^2CH-^1\overset{\oplus}{C}H_2}{}$$

These written structures indicate that there are two sites where electron supply will be formally received from a nucleophilic donor, namely at C_1 or C_3. A neutral water molecule is clearly the most likely nucleophile, but attack at C_3, which is in any case sterically unfavourable, will yield the initial compound. To yield the new product, attachment of water takes place at c_1 and the intermediate then deprotonates:

$$\underset{CH_3}{\overset{CH_3}{C}}=CH-\overset{\oplus}{C}H_2 \rightarrow \underset{CH_3}{\overset{CH_3}{C}}=CH-CH_2-\overset{\oplus}{O}H_2$$
$$\underset{:OH_2}{}$$

$$\downarrow -H^{\oplus}$$

$$\underset{CH_3}{\overset{CH_3}{C}}=CH-CH_2OH$$

In many reactions of allylic compounds a mixture of two isomeric products is obtained. This is because competition can take place for the two possible positions for nucleophilic attack on the intermediate carbonium ion. Thus the hydrolysis of crotyl chloride (essentially the replacement of —Cl by —OH), in aqueous acetone, yields approximately equal quantities of the normal and rearranged products:

$$CH_3—CH=CH—CH_2Cl \xrightarrow[\text{warm}]{H_2O} \begin{array}{l} CH_3—CH=CH—CH_2OH \text{ (normal)} \\ + \\ CH_3—CH—CH=CH_2 \text{ (rearranged)} \\ \qquad | \\ \qquad OH \end{array}$$

crotyl chloride

Here the resonance forms of the intermediate carbonium ion are:

$$CH_3—CH=CH\overset{\oplus}{—}CH_2 \leftrightarrow CH_3—\overset{\oplus}{CH}—CH=CH_2$$

the first leading to the normal substitution product, and the second to the rearranged substitution product.

Questions

1 The acetylenic Grignard reagent $CH\equiv C\overset{\delta-}{—}\overset{\delta+}{MgBr}$, prepared by the reaction sequence

$$C_2H_5Br+Mg \xrightarrow[\text{in dry ether}]{\text{reflux turnings}} C_2H_5MgBr \xrightarrow{CH\equiv CH} CH\equiv CMgBr+C_2H_6\uparrow$$

can be used to synthesize acetylenic tertiary alcohols by reaction with a ketone in dry ether, followed by decomposition of the intermediate addition compound with dilute acid. Write down the structure of the intermediate addition compound for acetone.

2 Write down the formulae of possible substances produced by the solvolysis (hydrolysis by the solvent) of the following compounds in absolute ethanol:

(a) $CH_3.CHCl.CH=CH_2$ (b) $CH_3.CH=CH.CH_2Cl$

3 Show clearly the complete reaction sequence for the hydrolysis with rearrangement of crotyl chloride outlined above. Suggest why —Cl is replaced by —OH so readily in this allylic alkyl halide.

4 Using reactions given so far in this book, outline a synthesis of crotyl chloride from acetaldehyde. Give the structural formula of a possible isomeric by-product.

5.4 The neopentyl rearrangement

Neopentyl alcohol (2,2-dimethylpropan–1–ol) can be conveniently prepared from t. butyl alcohol by a Grignard synthesis:

$$(CH_3)_3C-OH \xrightarrow[HCl]{conc.} (CH_3)_3C-Cl \xrightarrow[in\ dry\ ether]{Mg\ turnings} (CH_3)_3C-Mg-Cl$$

t. butanol   t. butyl chloride   t. butyl magnesium chloride (Grignard reagent)

Then

addition compound

$$CH_3-\underset{\underset{CH_3}{|}}{\overset{\overset{CH_3}{|}}{C}}-CH_2OH + MgCl_2$$

neopentyl alcohol

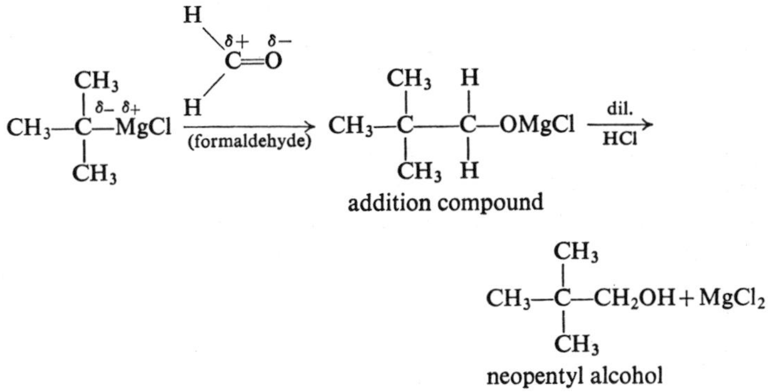

When this primary alcohol is treated with concentrated sulphuric acid, two isomeric olefins are produced by a net dehydration process:

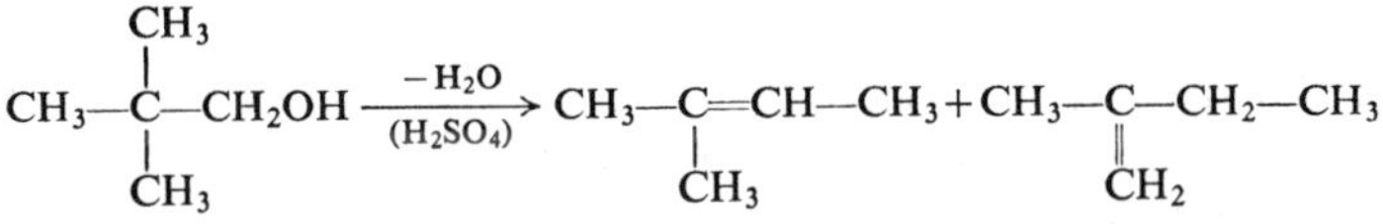

These olefins, which can be separated by preparative gas chromatography and distinguished by the different products formed with ozone (section 3.1E), must clearly arise by the rearrangement of the basic carbon skeleton:

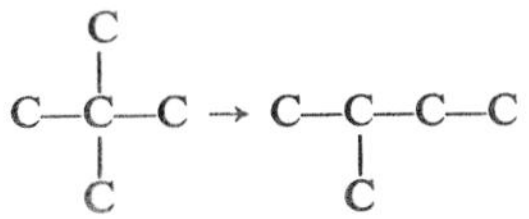

In accord with our previous descriptions of alcohol reactivity, the formal first stages of the reaction will be protonation of the hydroxyl group followed by loss of a water molecule to yield a carbonium ion:

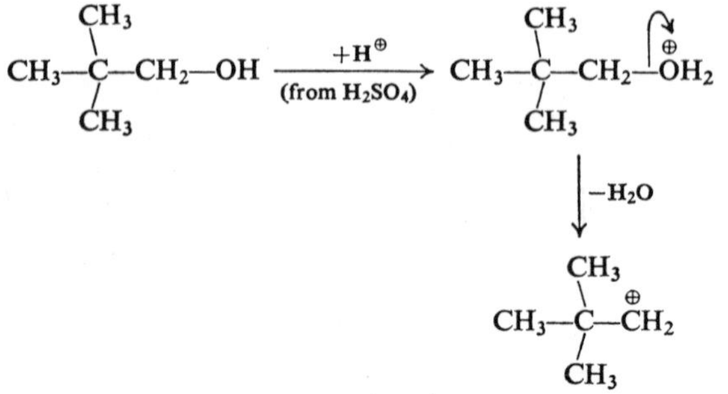

$$CH_3-\underset{\underset{CH_3}{|}}{\overset{\overset{CH_3}{|}}{C}}-CH_2-OH \xrightarrow[\text{(from }H_2SO_4)]{+H^\oplus} CH_3-\underset{\underset{CH_3}{|}}{\overset{\overset{CH_3}{|}}{C}}-CH_2-\overset{\oplus}{O}H_2$$

$$\downarrow -H_2O$$

$$CH_3-\underset{\underset{CH_3}{|}}{\overset{\overset{CH_3}{\backslash}}{C}}-\overset{\oplus}{C}H_2$$

In a substitution reaction, this carbonium ion would combine with a nucleophile to yield a stable product. In a normal elimination reaction, it would stabilize itself by deprotonation, but in this case is unable to do so because the adjacent carbon atom has no hydrogen atom to lose. Consequently, as this intermediate is a relatively unstable primary carbonium ion (see page 59), it will rearrange itself to a more stable tertiary carbonium ion by the migration of a methyl group *with its pair of bonding electrons:*

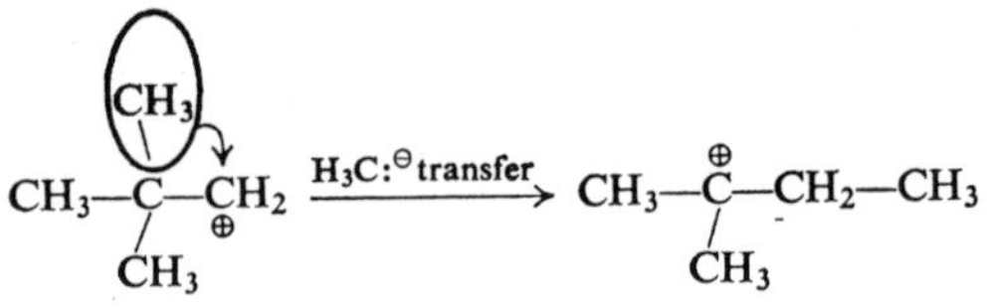

$$CH_3-\underset{\underset{CH_3}{|}}{\overset{\overset{\widehat{CH_3}}{\backslash}}{C}}-\overset{\oplus}{C}H_2 \xrightarrow{H_3C:^\ominus \text{ transfer}} CH_3-\underset{\underset{CH_3}{|}}{\overset{\oplus}{C}}-CH_2-CH_3$$

The tertiary carbonium ion is now able to deprotonate from either the methylene group or a methyl group adjacent to the electron deficient carbon atom to give the two olefinic products:

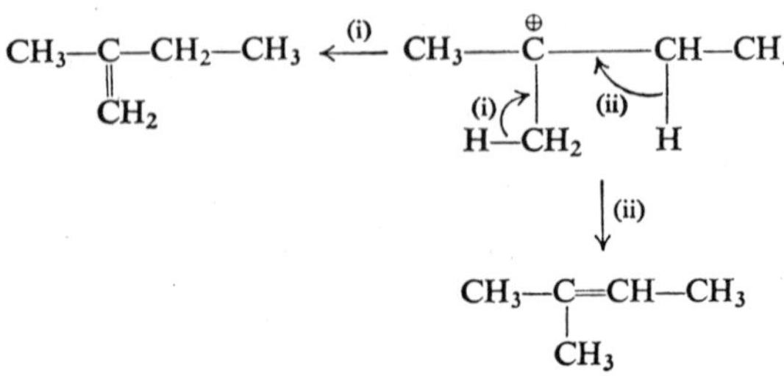

$$CH_3-\underset{\underset{CH_2}{\|}}{C}-CH_2-CH_3 \overset{(i)}{\longleftarrow} CH_3-\overset{\oplus}{C}\underset{\underset{H-CH_2}{\overset{(i)}{\nwarrow}}}{\overset{}{}}-CH-CH_3 \; \underset{H}{\overset{(ii)}{}}$$

$$\downarrow (ii)$$

$$CH_3-\underset{\underset{CH_3}{|}}{C}=CH-CH_3$$

Rearrangements of carbonium ions of suitable structure are widespread in organic reactions. Any reaction which is known to proceed through carbonium ion intermediates is liable to be complicated by rearrangements. Thus, in the Friedel-Crafts reaction of benzene with neopentyl chloride, the rearranged product predominates because the reaction proceeds through the carbonium ion of the alkyl halide (section 2.5A):

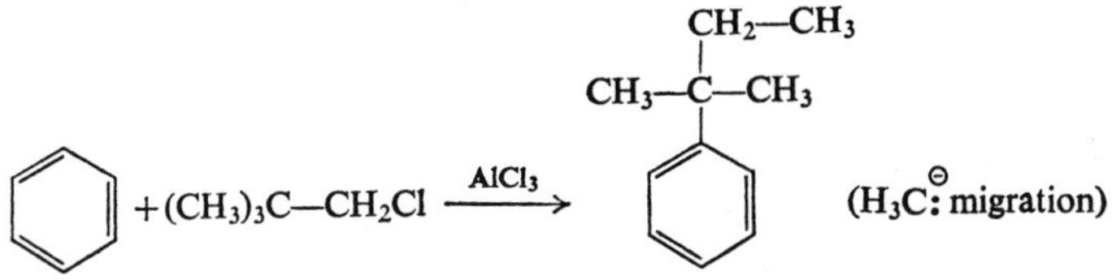

Similarly, when *n*.propylamine is treated with nitrous acid, the reaction proceeds by way of the primary carbonium ion $CH_3.CH_2.\overset{\oplus}{CH_2}$ (section 2.6A). This can rearrange to the more stable secondary carbonium ion by the effective migration of hydrogen with its pair of bonding electrons:

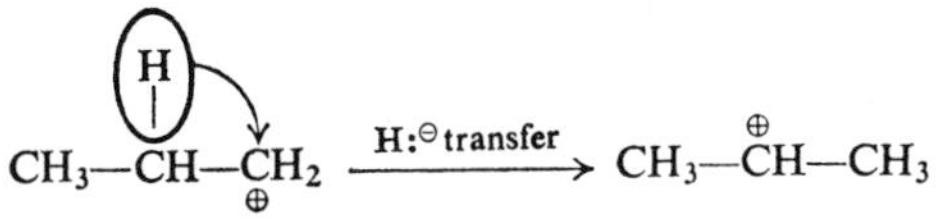

The predominant alcohol from the nitrous acid reaction is not therefore *n*.propanol, $CH_3.CH_2.CH_2OH$, but isopropanol, $CH_3.CH(OH).CH_3$. Propylene, CH_3—CH=CH_2, is also obtained in this reaction by deprotonation of either the primary or the secondary carbonium ion shown above.

Questions

1 State three general ways in which a carbonium ion consisting of three or more carbon atoms can react.

2 Outline the synthesis of *t*.butanol from acetone.

3 Give the formulae of the predominant products formed when benzene is treated in the presence of aluminium chloride with
   (a) *n*. propyl chloride
   (b) isopropyl chloride
   (c) $(CH_3)_3C.CH_2.CH_2Cl$.

4 Explain why the acid catalyzed dehydration of the compound $(CH_3)_3C.CH(OH).CH_3$ gives rise to only a trace of the olefin $(CH_3)_3C.CH=CH_2$. Give the formula of the major product.

5.5 The cumene hydroperoxide process—

A MODERN METHOD FOR THE INDUSTRIAL SYNTHESIS OF PHENOL

A new method has been introduced in recent years for the commercial synthesis of phenol from benzene in three main stages:

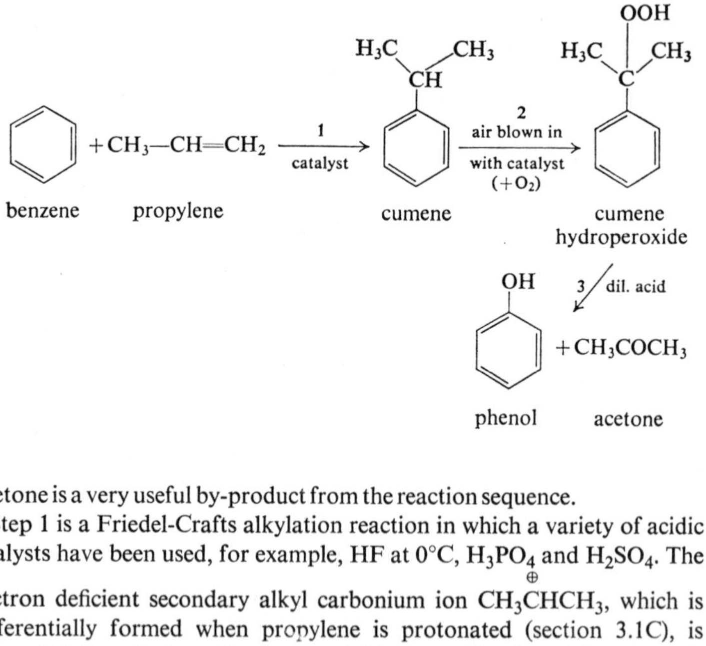

benzene   propylene   cumene   cumene
                                                      hydroperoxide

phenol   acetone

Acetone is a very useful by-product from the reaction sequence.

Step 1 is a Friedel-Crafts alkylation reaction in which a variety of acidic catalysts have been used, for example, HF at 0°C, H_3PO_4 and H_2SO_4. The electron deficient secondary alkyl carbonium ion $CH_3\overset{\oplus}{C}HCH_3$, which is preferentially formed when propylene is protonated (section 3.1C), is responsible for attack on the electron rich benzene ring. Step 2 is an autoxidation reaction and takes place in the presence of a base, for example, NaOH, or peroxides. This process may be ionic or free-radical in nature, but its essential feature is that a hydrogen atom is abstracted from a tertiary benzylic carbon atom and a carbon–oxygen bond is produced. The free radical self propagating chain mechanism which occurs in the presence of peroxides can be represented:

R·
Radical
initiating
catalyst
$+ Ph-C(CH_3)(H)(CH_3) \rightarrow R:H + Ph-C·(CH_3)(CH_3)$

$+O_2$

Ph—CH(CH_3)(CH_3)

Ph—C·(CH_3)(CH_3) $+$ Ph—C—O—OH(CH_3)(CH_3) $\longleftarrow$ Ph—C—O—O·(CH_3)(CH_3)

etc., etc. $(Ph = C_6H_5)$

Stage 3 of the reaction sequence is the acid catalyzed rearrangement step. The hydroperoxide which contains an —OH group is first protonated in the acidic medium. The weakened O—O bond can then break to yield an oxycation, water being eliminated.

$$CH_3-\underset{Ph}{\underset{|}{C}}-O-O-H \xrightarrow{H^\oplus} CH_3-\underset{Ph}{\underset{|}{C}}-O-\overset{\oplus}{O}\underset{H}{\overset{H}{\big\langle}} \rightarrow CH_3-\underset{Ph}{\underset{|}{C}}-\overset{..}{\underset{}{O}}{}^\oplus + H_2O$$

The oxygen atom retained in the molecule is electron deficient, having but six valence electrons. This is the vital factor in the total reaction; the positively charged oxygen atom will attract any available nucleophile. However, in aqueous acid solution, there is a lack of any nucleophile which will give rise to a stable product. Therefore, the electron deficiency of the oxygen atom is satisfied by migration of the phenyl group *with its pair of bonding electrons* on to the oxygen atom. This produces a more stable carbonium ion, this being the driving force for the migration.

$$CH_3-\underset{\underset{\oplus}{\big(Ph\big)}}{\overset{|}{C}}-\overset{..}{O}: \xrightarrow{Ph:^\ominus \text{ migration}} CH_3-\underset{\oplus}{\underset{|}{C}}-\overset{..}{O}-Ph$$

The phenyl group migrates rather than a methyl group because it is a more powerful electron donor, possessing the 'excess' π electron cloud.

It should be emphasised, as in earlier discussions, that a reaction formulation in which the step of dissociation of H_2O from a protonated hydroxylic entity is shown as a discrete process, is a convenience. The intermediate cation may not even have a real existence, as a simultaneous elimination of water and migration (or substitution in suitable circumstances) is equally possible:

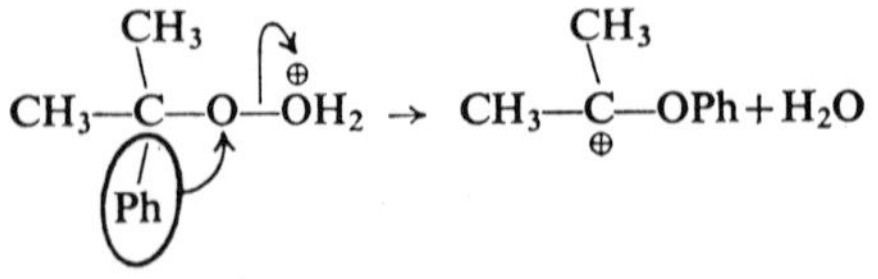

$$CH_3-\overset{\overset{\displaystyle CH_3}{|}}{\underset{(Ph)}{C}}-O-\overset{\oplus}{O}H_2 \rightarrow CH_3-\overset{\overset{\displaystyle CH_3}{|}}{\underset{\oplus}{C}}-OPh + H_2O$$

The writing of the equation for the reaction with the extra oxy-cation as a discrete intermediate illustrates the mechanistic pathway more clearly.

The more stable carbonium ion will not have a long lifetime in the solution, and it will readily pick up a molecule of water from the aqueous solution. If this entity deprotonates, it yields the acid sensitive hemiketal (section 3.2B) of phenol and acetone:

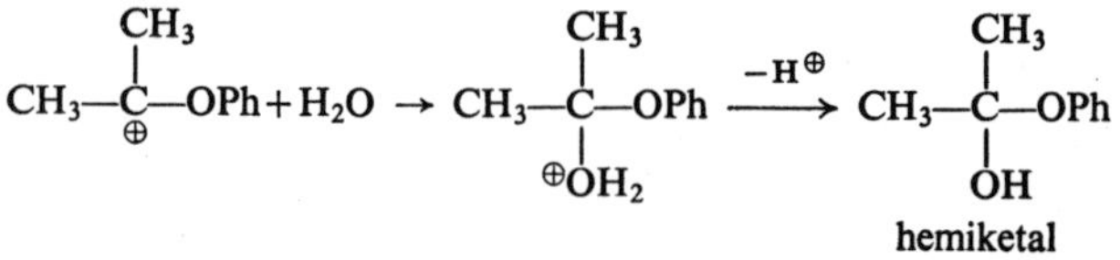

$$CH_3-\overset{\overset{\displaystyle CH_3}{|}}{\underset{\oplus}{C}}-OPh + H_2O \rightarrow CH_3-\overset{\overset{\displaystyle CH_3}{|}}{\underset{\overset{\displaystyle \oplus OH_2}{}}{C}}-OPh \xrightarrow{-H^\oplus} CH_3-\overset{\overset{\displaystyle CH_3}{|}}{\underset{\overset{\displaystyle OH}{}}{C}}-OPh$$
hemiketal

The hemiketal will quickly yield the final products by a sequence of the type:

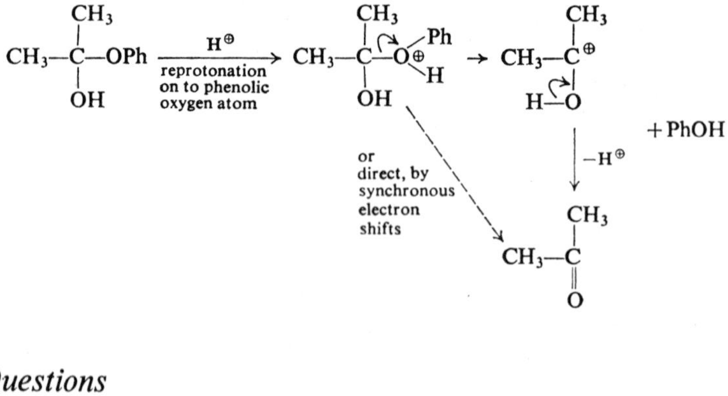

$$CH_3-\overset{\overset{\displaystyle CH_3}{|}}{\underset{\overset{\displaystyle OH}{}}{C}}-OPh \xrightarrow[\substack{\text{reprotonation}\\ \text{on to phenolic}\\ \text{oxygen atom}}]{H^\oplus} CH_3-\overset{\overset{\displaystyle CH_3}{|}}{\underset{\overset{\displaystyle OH}{}}{C}}-\overset{Ph}{\underset{H}{O^\oplus}} \rightarrow CH_3-\overset{\overset{\displaystyle CH_3}{|}}{\underset{\overset{\displaystyle H-O}{}}{C^\oplus}}$$

+ PhOH

or direct, by synchronous electron shifts

$-H^\oplus$

$$CH_3-\overset{\overset{\displaystyle CH_3}{|}}{\underset{\overset{\displaystyle O}{\|}}{C}}$$

Questions

1 Draw clearly the synchronous electron shifts referred to by the dotted arrow above.

2 Which group would be most likely to migrate with its pair of bonding electrons if the hydroperoxide

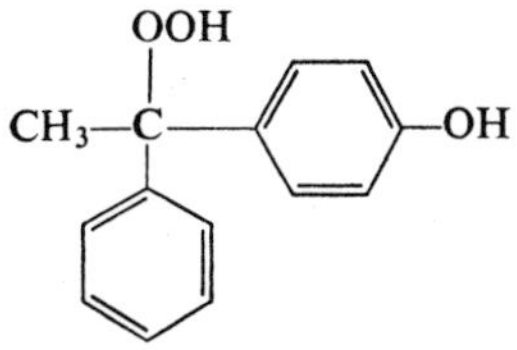

were treated with acid? Give your reason and write down the formulae of the final products.

SUMMARY

Functional group migration, or a change in the basic carbon skeleton are involved in molecular rearrangements. Carbonium ions frequently play a part in skeletal changes, where migration of a group of atoms with a pair of bonding electrons takes place to an electron deficient centre to yield a more stable entity: this type of rearrangement is accelerated if the migrating group is potentially strongly electron releasing.

Miscellaneous Exercises

1 Acetylenes add water across their triple bond in the presence of aqueous $H_2SO_4/HgSO_4$ as catalyst. Give the formulae of the compounds produced when (a) acetylene (b) methyl acetylene are so treated. Which generalization helps you to predict the product in (b)?

2 Cuprous compounds catalyze substitutions in benzene diazonium chloride. Give the formulae of compounds obtained when benzene diazonium chloride is treated with (a) Cu_2Cl_2/HCl (b) $Cu_2(CN)_2/KCN$.

3 Suggest why benzaldehyde, C_6H_5—CHO, is slightly less reactive than acetaldehyde towards nucleophiles.

4 Electrolysis of aqueous sodium acetate yields ethane and carbon dioxide at the anode (1:2 by volume). The process involves free radical intermediates. Indicate the reaction path.

5 Complete the following sentence:
'The .... induced pull of electrons is of .... range, whereas resonance effects which involve .... overlap are of .... range.'

6 The C—Cl bond of ethyl chloride has a length of 1·76 Å and the $> C=C <$ bond of ethylene is 1·34 Å long. Explain the following bond lengths for vinyl chloride (chloroethylene):

   $CH_2=CH$—Cl
   1·38Å 1·69Å

and suggest why vinyl chloride is less readily hydrolyzed (—Cl → —OH) than ethyl chloride by aqueous alkali.

7 Give two permissible electronic resonance structures for each of the following: (a) vinyl chloride (b) methyl formate (c) acetaldehyde.

8 Methyl groups attached to an aromatic ring can be directly and selectively oxidized to carboxyl groups using $KMnO_4$. Outline a synthesis of 2,4,6 trinitrobenzoic acid from benzene.

9 Suggest why the heat of hydrogenation of 1,3 cyclohexadiene (page 29) is some 5 kcal/mole less than that of 1,4 cyclohexadiene.

10 The pK_a value of $CH_3COCH_2COOC_2H_5$ is about 10. Explain.

11 By means of a labelled diagram, indicate the electronic effects which operate in amides, $RCONH_2$, which explain why the carbonyl group is inert to addition reactions.

12 Give the formulae of the olefins obtained when the following primary amines are treated with nitrous acid: (a) ethylamine (b) n-butylamine (c) neopentylamine.

Answers to Questions

1 page 6
1 (a) ions
  (b) provide; electrophilic; base

2

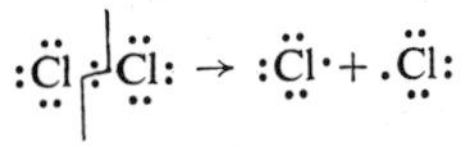

$$:\overset{..}{\underset{..}{Cl}}\!\!\!-\!\!\!\overset{|}{\underset{|}{Cl}}: \;\rightarrow\; :\overset{..}{\underset{..}{Cl}}\!\cdot + \cdot\overset{..}{\underset{..}{Cl}}:$$

Radicals are electrically neutral.

2.1 page 9
If there is excess ethanol in the reaction mixture, then there is a far higher probability of a neutral ethanol molecule acting as nucleophile on protonated ethanol to yield diethyl ether:

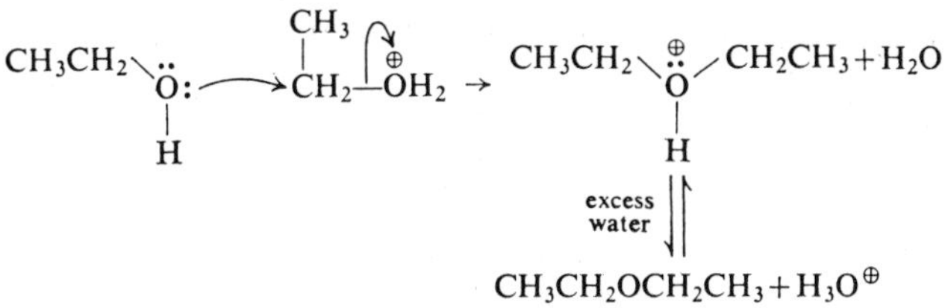

$$CH_3CH_2OCH_2CH_3 + H_3O^{\oplus}$$

The last stage of deprotonation is favoured by the addition of excess water after the main stage of the reaction is completed. This drives the equilibrium in favour of diethyl ether, the water molecule being a ready proton acceptor.

2.2 page 14
1 The nucleophile is hydroxyl ion, and ethyl alcohol is the chief product.

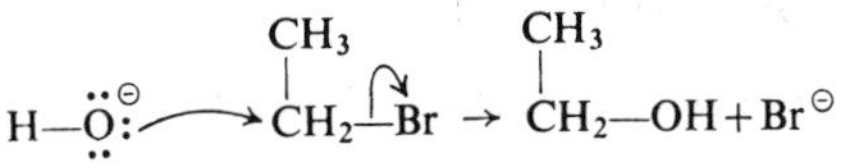

2 The nucleophile is cyanide ion, and ethyl cyanide is the major product.

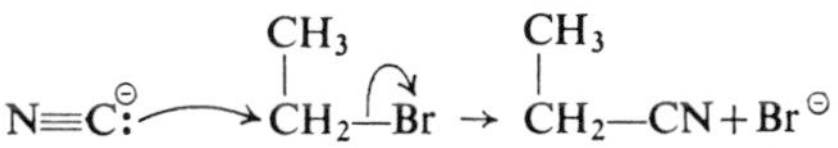

3 Acidity order (strongest first):

  C, A, D, E, B, F

4 The following have dipole moments:
  hydrogen chloride, chloroform, chlorobenzene, *cis*-dichloroethylene.

2.3 page 22
1 Three steps:
  1. Oxidation to acetaldehyde $CH_3.CHO$
  2. Iodination to tri-iodoacetaldehyde CI_3CHO
  3. Hydrolysis to iodoform, and sodium formate, $HCOO^{\ominus}Na^{\oplus}$.

104

2 Iodoform given by compounds A, D and E.

3 Bubble chlorine into a warm solution of acetone in dilute sodium carbonate or sodium hydroxide. The chloroform sinks to the bottom of the mixture (density $1 \cdot 5$ gm/cm³). Alternatively, aqueous acetone can be added dropwise to a hot solution of bleaching powder (or sodium hypochlorite), and the chloroform steam distilled from the reaction mixture as it is formed.

2.4 page 26

1

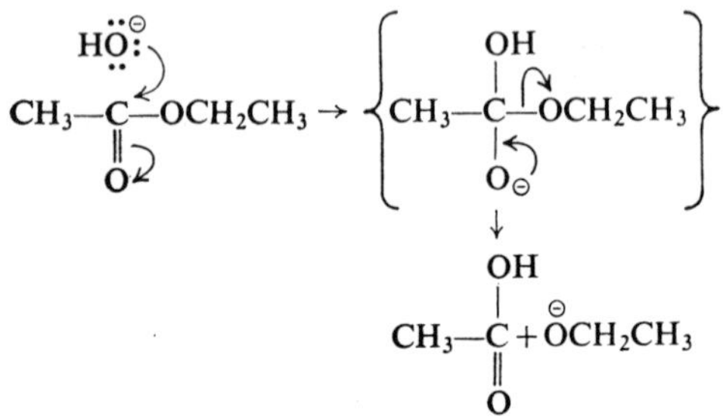

This nucleophilic attack by hydroxyl ion is followed by a series of proton exchanges to yield the final products:

$$CH_3COOH + OH^\ominus \rightarrow CH_3COO^\ominus + H_2O$$

$$H_2O + {}^\ominus OCH_2CH_3 \rightarrow OH^\ominus + HOCH_2CH_3$$

2 Again, this reaction is one of nucleophilic attack followed by proton exchange:

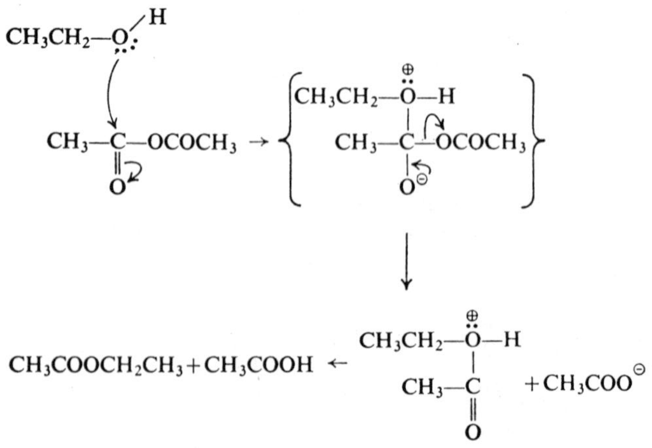

2.5 page 30

1 The carbon atoms of graphite are sp^2 hybridized similar to those of benzene, and there are 'surplus' π electrons above and below the plane of each sheet of carbon atoms. These 'delocalized' electrons are not attached to any particular carbon atom and their mobility accounts for graphite's conducting properties.

2 Yes. The cyclohexadienes have a negligible resonance energy and hence exhibit the normal reactions of olefinic compounds.

3 The π electron cloud of the aromatic ring acts as a shield to the nucleophilic hydroxyl ion which is the attacking entity in the hydrolysis reaction, thus making drastic conditions necessary.

2.5A page 32

1 Product is $C_6H_5COCH_3$—acetophenone, an aromatic ketone. Attacking electrophile is $CH_3CO^\oplus$.

2 The chlorine will react with the iron to form ferric chloride, and this can then react as a Lewis acid with more chlorine to yield the attacking $Cl^\oplus$ ion:

$$FeCl_3 + Cl\!-\!Cl \rightarrow FeCl_4^\ominus + Cl^\oplus$$

Ferric chloride is therefore the true catalyst for the reaction.

2.5B page 36

Product is

$$\overset{\oplus}{N}(CH_3)_3 I^\ominus$$

and the rate of nitration is slower than for benzene. The positive charge on the 4-covalent nitrogen atom attracts electrons from the aromatic ring preferentially from the $o.$ and $p.$ positions to leave the ring less prone to electrophilic attack by the $NO_2^\oplus$ ion.

2.5C page 39

1 The rate of nitration of phenol will be considerably faster than that of benzene. With excess nitrating solution, the electrophilic $NO_2^\oplus$ ion will react at the positions of highest electron density (ortho and para) to yield 2, 4, 6 trinitrophenol (picric acid):

Mononitration, using less vigorous conditions (dil. HNO_3) will yield a mixture of ortho and para nitrophenols:

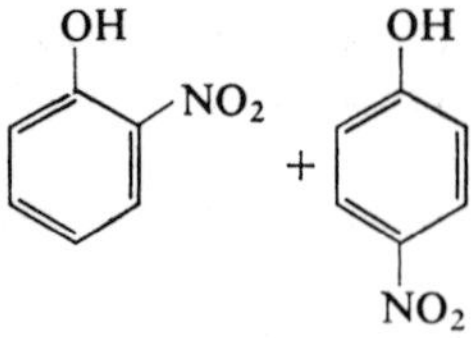

2 The lone pair of electrons on the nitrogen atom make the electron distribution of aniline analogous to that of phenol. Electron donation by the $-NH_2$ group to the ring makes the ortho and para positions in aniline more negative than in benzene. Bromination is therefore rapid, yielding, 2, 4, 6 tribromoaniline.

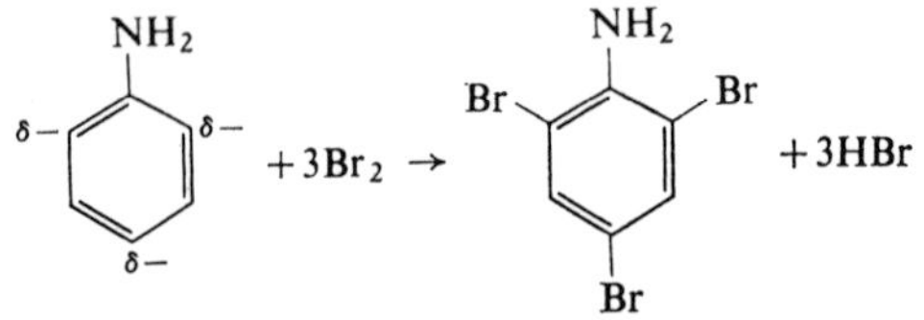

3 Methylamine is a stronger base ($pK_b 3\cdot4$) than aniline ($pK_b 9\cdot4$). The lone electron pair on nitrogen in the latter compound is less available for coordinating a proton, due to its tendency to participate in the resonance of the aromatic ring. Protonation of aniline would require localization of the electron pair on nitrogen and loss of the additional resonance stabilization from contributions such as

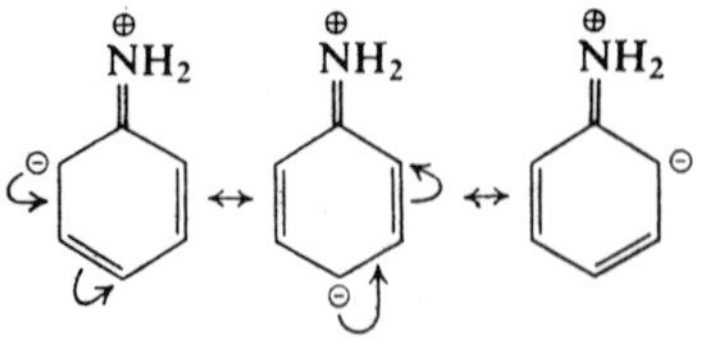

Proton acceptance by aniline is therefore less favoured than for methylamine where no such resonance stabilization occurs.

2.6 page 47

1.

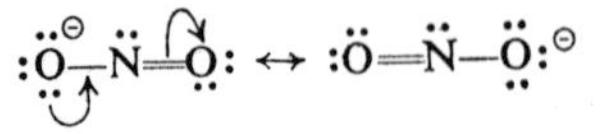

2. $[N_2O_3]$

3 Ortho-nitroaniline (pK_b 14·3) is a considerably weaker base than aniline because the electron drawing $-NO_2$ group in the ring results in a very efficient electron donation to the ring by the lone pair on the amino nitrogen atom; this makes this electron pair much less available for proton coordination.

4 The methyl carbonium ion formed by the usual pathway would react with a methanol molecule as nucleophile to yield, after deprotonation, dimethyl ether:

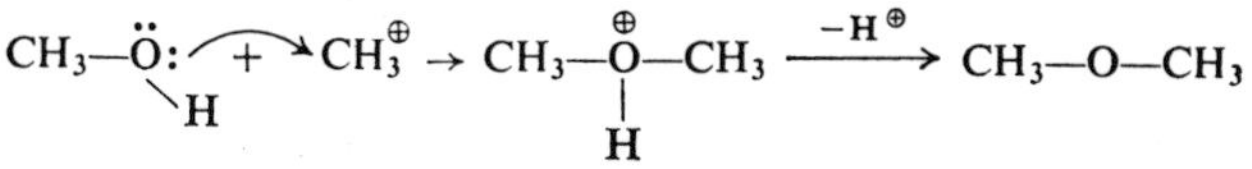

$$CH_3-\overset{..}{\underset{\underset{H}{|}}{O}}: \overset{\frown}{\quad} + {}^{\searrow}CH_3^{\oplus} \rightarrow CH_3-\overset{\oplus}{\underset{\underset{H}{|}}{O}}-CH_3 \xrightarrow{-H^{\oplus}} CH_3-O-CH_3$$

5 Iodobenzene, C_6H_5-I, would be the major product formed by nucleophilic attack by $I^{\ominus}$ on the intermediate $C_6H_5^{\oplus}$ carbonium ion.

2.7 page 51

(b) 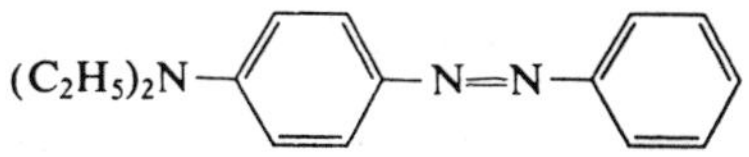 $(C_2H_5)_2N$—⟨benzene ring⟩—N=N—⟨benzene ring⟩

(d) 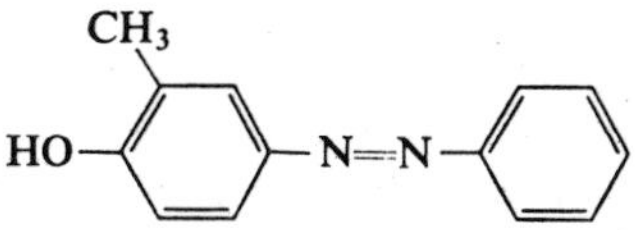 HO—⟨benzene ring with CH₃⟩—N=N—⟨benzene ring⟩

$$CH_3$$

Benzene (unactivated) and nitrobenzene (deactivated) fail to react.

3.1A page 55
1 $Br-CH_2-CH_2-Br$; $Br-CH_2-CH_2-OH$; $Br-CH_2-CH_2-Cl$.
2 *Cis* 2-butene yields a resolvable dl mixture.

3.1B page 56
1 5
2 Propane, $CH_3-CH_2-CH_3$
3 *Cis* 2-butene

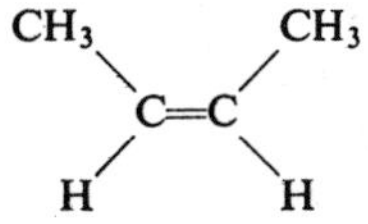

$$\underset{H}{\overset{CH_3}{\diagdown}}C=C\underset{H}{\overset{CH_3}{\diagup}}$$

3.1C page 61
1 CH_3-CHBr_2. Ionic mechanism.

2 $CH_3-\underset{\underset{OH}{|}}{CH}-\underset{\underset{Cl}{|}}{CH_2}$

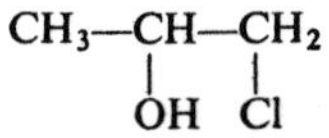

3.

$$CH_3\text{—}\overset{\delta+}{CH}\text{—}\overset{\delta-}{CH_2}\overset{\curvearrowleft}{H^\oplus} \xrightarrow[\text{ionized acid}]{\substack{\text{electrophilic attack} \\ \text{by } H^\oplus \text{ from}}} CH_3\text{—}\overset{\oplus}{CH}\text{—}CH_3 \xrightarrow[\substack{\text{nucleophilic} \\ \text{attack by} \\ H_2O}]{H\diagdown \ddot{O}\diagup H} CH_3\text{—}\underset{CH_3}{\overset{\oplus}{CH}\text{—}OH_2}$$

$$\downarrow -H^\oplus$$

$$CH_3\text{—}\underset{CH_3}{CH}\text{—}OH$$

Product is propan-2-ol (isopropyl alcohol) by Markovnikov addition.

4    $CH_3\text{—}\underset{\delta+}{CH}=\underset{\delta-}{CH_2} \xrightarrow{Br\cdot} CH_3\text{—}\overset{\cdot}{CH}\text{—}CH_2Br \xrightarrow{H\cdot} CH_3\text{—}CH_2\text{—}CH_2Br$

The bromine atom attacks the propylene molecule at the centre of highest electron density *before* the hydrogen atom, probably because it is more electronegative (more electron seeking) than hydrogen. If the order of attack were reversed, 2-bromopropane would result because the more stable secondary intermediate radicals are favoured.

5 The addition is a *trans* addition process.

3.1D   page 64

1 False. Ethylene glycol is not optically active, and although the reactions will proceed by the *cis* and *trans* routes, once formed, the products will be identical due to the easy rotation that is possible by normal thermal molecular motions about the carbon–carbon bond.

2 Both reagents form tartaric acids

$$CH_3\text{—}CH(OH)\text{—}CH(OH)\text{—}CH_3$$
$$\underset{COOH}{|} \qquad \underset{COOH}{|}$$

(a) OsO_4 yields meso tartaric acid (non resolvable).

(b) CH_3CO_3H yields racemic dl tartaric acid (resolvable).

3 The formation of the resonance stabilized acetate anion favours the indicated ionisation. The electrophilic $^\oplus OH$ ion will attack the electron rich olefin.

Some

$$CH_3COO\text{—}\underset{\overset{|}{CH_2\text{—}OH}}{CH_2}$$

will be formed by rear attack of acetate anion on the protonated epoxide.

3.1E page 65

1 (a) acetaldehyde (2 molecules).
   (b) acetone and formaldehyde.
   (c) methyl ethyl ketone and formaldehyde.

2 Cyclic olefins (for example, cyclohexene).

3.2A page 68

1 (a) $CH_3-CH_2-CH_2-CH_2-COOH$
   (b) $CH_3-CH=CH-CH_2-CH_2OH$

2 (a) CH_3-CH_2OH
   (b) $CH_3-CH_2OH + CH_3OH$

3.2B page 70

1

$$\underset{H}{\overset{CH_3}{>}}C\overset{OH}{\underset{\overset{..}{O}CH_3}{<}} + H^\oplus \rightarrow \underset{H}{\overset{CH_3}{>}}C\overset{O-H}{\underset{\overset{\oplus}{O}}{<}} \underset{CH_3}{\overset{H}{}}$$

$$\underset{H}{\overset{CH_3}{>}}C=O + CH_3OH + H^\oplus$$

2 Esters are most readily hydrolyzed with aqueous alkali. Aldehydes very often undergo complex polymerization reactions in the presence of $OH^\ominus$ ions. Hence the aldehyde-ester must be protected at the aldehyde group, most conveniently with a dimethyl acetal:

$$\underset{CH_2-OCOCH_3}{CH_2-CHO} \xrightarrow[\text{dry HCl}]{CH_3OH} \underset{CH_2-OCOCH_3}{CH_2-CH(OCH_3)_2} \xrightarrow{NaOH} \underset{CH_2-OH}{CH_2-CH(OCH_3)_2}$$

$$\xrightarrow[\text{aq. HCl}]{\text{dilute}} \underset{CH_2-OH}{CH_2-CHO}$$

Some ester hydrolysis may occur during the first stage of this sequence.

3.2C page 71

1
$$\underset{H}{\overset{CH_3}{>}}\overset{\delta+}{C}\overset{\delta-}{=}O \quad \overset{\ominus}{:}SO_3H \xrightarrow[\text{attack}]{\text{nucleophilic}} \underset{H}{\overset{CH_3}{>}}C\overset{O^\ominus}{\underset{SO_3H}{<}} \xrightarrow[\text{transfer}]{\text{proton}} \underset{H}{\overset{CH_3}{>}}C\overset{OH}{\underset{SO_3^\ominus Na^\oplus}{<}}$$

2
$$\underset{H}{\overset{CH_3}{>}}\overset{\delta+}{C}\overset{\delta-}{=}O \quad :NH_3 \xrightarrow[\text{attack}]{\text{nucleophilic}} \underset{H}{\overset{CH_3}{>}}C\overset{O^\ominus}{\underset{\overset{\oplus}{N}H_3}{<}} \xrightarrow[\text{transfer}]{\text{proton}} \underset{H}{\overset{CH_3}{>}}C\overset{OH}{\underset{NH_2}{<}}$$

3 Esters (and other carboxylic acid derivatives) do not undergo the simple addition reactions of aldehydes and ketones because the carbonyl carbon atom is less positive due to the operation of the resonance effect.

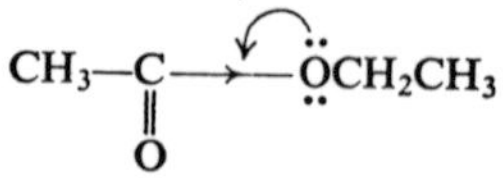

$$CH_3-C \xrightarrow{} \ddot{O}CH_2CH_3$$
$$\overset{\|}{O}$$

This atom is therefore not prone to attack by relatively weak nucleophiles like $CN^{\ominus}$.

3.2D page 72

1 (a) $\overset{\ominus}{C}H_2-CO-CH_3$

(b) The electron donating effect of the extra methyl group in acetone makes the carbonyl carbon atom less positive, and hence less susceptible to nucleophilic attack, than in acetaldehyde.

2 Aldol

3
$$CH_3CHO + HCHO \xrightarrow{\text{trace } OH^{\ominus}} HOCH_2-CH_2CHO$$
$$\text{(1 molecule)}$$

Reaction proceeds by attack of $\overset{\ominus}{C}H_2-CHO$ on formaldehyde carbonyl carbon atom.

3.3 page 76

1 (a) $CH_3-CH_2-CH_2-CH_2-NH_2$ (b) $CH_3-CH=CH-CH_2-NH_2$

2

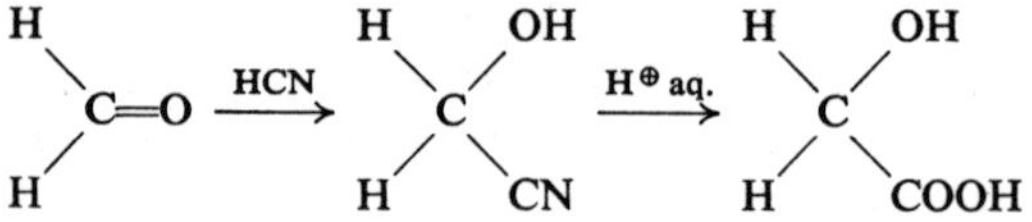

3

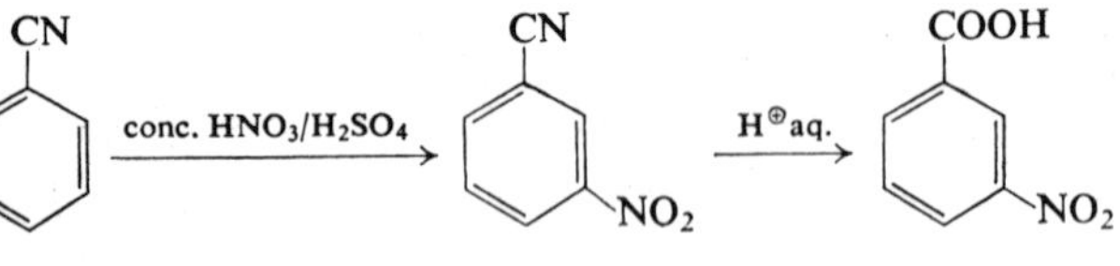

4.1 page 79

1

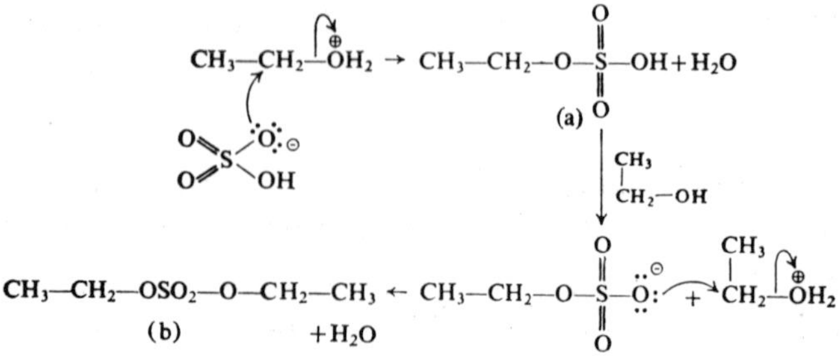

(a) by nucleophilic attack by bisulphate ion on protonated ethanol.

(b) by nucleophilic attack by $C_2H_5OSO_2O^\ominus$ (formed by a proton exchange reaction) on protonated ethanol.

(c) by nucleophilic attack by ethanol on protonated ethanol, followed by deprotonation:

$$CH_3CH_2 \overset{\cdot\cdot}{\underset{H}{O}}: \quad \overset{CH_3}{\underset{}{\mid}} CH_2 \overset{\oplus}{-OH_2} \xrightarrow{-H_2O} CH_3CH_2 \overset{\oplus}{\underset{H}{-O}} -CH_2CH_3$$

$$\downarrow -H^\oplus$$

$$CH_3CH_2OCH_2CH_3$$

(c)

2 Diethyl ether formed by an external or intermolecular dehydration: that is, two ethanol molecules shed one molecule of water.

$$CH_3CH_2 \dashv OH + H \vdash OCH_2CH_3 \xrightarrow{-H_2O} CH_3CH_2-O-CH_2CH_3$$

3

$$\begin{array}{c} CH_2 \\ \parallel \\ CH_2 \end{array} H^\oplus \xrightarrow{1} \begin{array}{c} \overset{\oplus}{CH_2} \leftarrow :OH_2 \\ \parallel \\ CH_2-H \end{array} \xrightarrow{2} \begin{array}{c} \overset{\oplus}{CH_2-OH_2} \\ \mid \\ CH_3 \end{array} \xrightarrow[-H^\oplus]{3} \begin{array}{c} CH_2-OH \\ \mid \\ CH_3 \end{array}$$

1. Electrophilic addition of $H^\oplus$ on to π electrons.
2. Nucleophilic attack by water on to carbonium ion.
3. Deprotonation.

4.2 page 81

$(CH_3)_2C{=}CHCOCH_3$

2 The two π electron systems of butadiene tend to overlap to give the central carbon–carbon bond some double bond character, and hence a shorter bond length than would be apparent from the written structure.

3 Nitrogen becomes attached to carbon by nucleophilic attack by the lone electron pair of nitrogen on the partially positive carbonyl carbon atom.

$(CH_3)_2C{=}N-NHC_6H_5$

4.3 page 84

1 $C_2H_5O^\ominus$

2 $CH_3.CH(OH).CH_3$ and $CH_3.CH(OC_2H_5).CH_3$ by substitution of $OH^\ominus$ or $C_2H_5O^\ominus$.

3

$$HO^\ominus\!\!\curvearrowright\!H \quad \overset{H}{\underset{Br}{\underset{\downarrow}{}}} \quad HO_2C \overset{}{\blacktriangleright} C \overset{}{-} C \cdots CO_2H \rightarrow H_2O + \overset{HO_2C}{\underset{Br}{}}C{=}C\overset{H}{\underset{CO_2H}{}} + Br^\ominus$$

4

$$HO:^{\ominus}\curvearrowright H$$
$$\underset{|}{CH_2}\text{—}CH_2\text{—}\overset{\oplus}{N}(CH_3)_3I^{\ominus} \;\rightarrow\; H_2O+CH_2\!\!=\!\!CH_2+N(CH_3)_3+I^{\ominus}$$

4.4 page 85

1 Add bromine to form the dibromo derivative of the olefin; distil out the unchanged paraffin at *ca*. 50°C, and boil the residue with Zn/C_2H_5OH to regenerate the olefin.

2 *Cis* 2-butene.

5.1 page 88

1 The mechanism of the Hofmann reaction requires deprotonation from nitrogen of an N-bromoamide under the influence of base. The N-methyl substituted amide will yield $CH_3CONBrCH_3$ as the final product because this compound cannot undergo the next step of deprotonation from nitrogen as the mechanism requires.

2 *p*-nitrobenzamide forms its amine slower than benzamide because the nitro group is an electron withdrawing deactivating substituent in the aromatic ring. The migrating group $-C_6H_4-NO_2$ is therefore a weaker electron donor than $-C_6H_5$.

5.2 page 91

1 $C_6H_5-NH_2$

$$CH_3-\underset{\underset{O}{\|}}{C}-Cl \;\rightarrow\; \left\{ \begin{array}{cc} C_6H_5-\overset{\oplus}{N}H_2 \\ CH_3-\underset{\underset{O^{\ominus}}{|}}{C}-Cl \end{array} \right\} \;\rightarrow\; \begin{array}{c} C_6H_5-\overset{\oplus}{N}H_2 \\ CH_3-\underset{\underset{O+Cl^{\ominus}}{\|}}{C} \end{array} \xrightarrow{-H^{\oplus}} \begin{array}{c} C_6H_5-NH \\ CH_3-\underset{\underset{O}{\|}}{C} \end{array}$$

2 Release of the Cl atom on to $I^{\ominus}$ yields I—Cl. The iodine atom is the more positive part of this interhalogen molecule (lower electronegativity coefficient), and hence it will be the electrophilic substituent into the aromatic ring, entering as $I^{\oplus}$. The ortho isomer will not be favoured mainly due to spatial interaction of the large iodine atom with the —NHCOCH$_3$ group, which would be adjacent to it at the ortho position. For similar reasons, para bromoacetanilide will be the major product when N-chloroacetanilide is treated with aqueous HBr.

3 Chlorine would be evolved to leave a residue of 2,4,6-trichloroacetanilide.

5.3 page 94

1

$$\begin{array}{ccc} CH_3 & & OMgBr \\ & \diagdown\;\diagup & \\ & C & \\ & \diagup\;\diagdown & \\ CH_3 & & C\!\equiv\!CH \end{array}$$

2 (a) 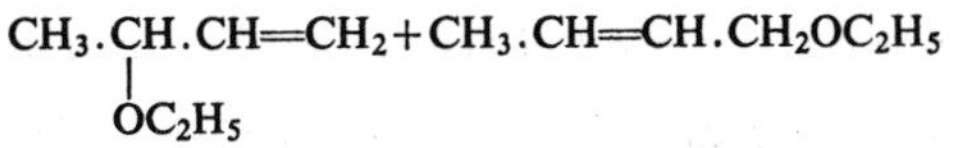 $CH_3.CH.CH=CH_2 + CH_3.CH=CH.CH_2OC_2H_5$
      |
      OC_2H_5

(b) As (a). Both compounds yield the same carbonium ion.

3 $CH_3-CH=CH-CH_2Cl \xrightarrow{-Cl^\ominus} CH_3-CH\overset{\frown}{=}CH\overset{\oplus}{-CH_2}$

$\updownarrow$

$CH_3-CH-CH=CH_2 \xleftarrow{H_2O} CH_3-\overset{\oplus}{CH}-CH=CH_2$
        |
      $_\oplus OH_2$

$\downarrow -H^\oplus$

$CH_3-CH-CH=CH_2$
        |
        OH

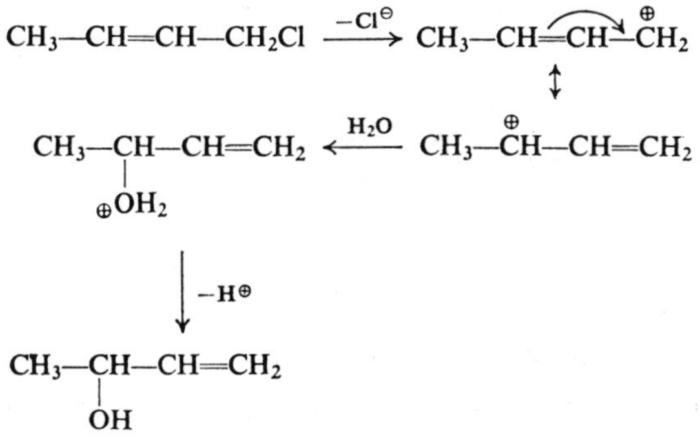

The ionization of $Cl^\ominus$ is favoured, and the reaction made easy, due to the resonance stabilization of the resultant carbonium ion.

4 $2CH_3CHO \xrightarrow{trace\ OH^\ominus} CH_3CH.CH_2.CHO \xrightarrow[-H_2O]{warm} CH_3CH=CH-CHO$
                                    |
                                    OH $LiAlH_4 \downarrow$

$CH_3CH=CH-CH_2Cl \xleftarrow[medium\ conc.]{HCl\ aq.} CH_3CH=CH-CH_2OH$

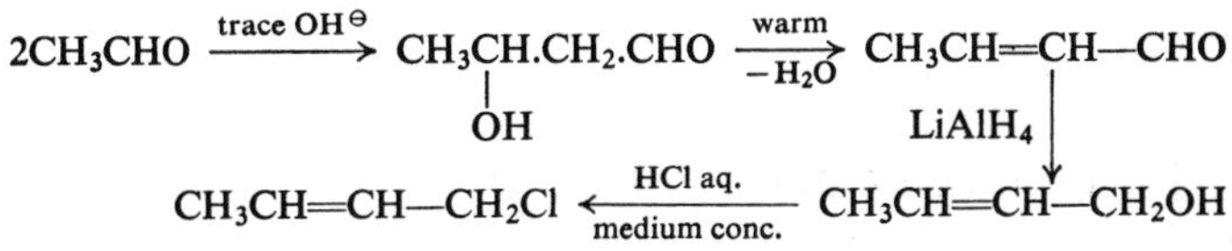

Possible isomeric by-product from the last step: $CH_3.CHCl.CH=CH_2$.

5.4 page 97

1 Carbonium ions can (a) combine with a nucleophile (b) eliminate a proton or (c) rearrange their structure.

2
$CH_3 \backslash$
$\qquad C=O + CH_3MgBr \xrightarrow{ether}$
$CH_3 /$

$CH_3 \quad OMgBr$
$\quad \backslash \ /$
$\qquad C \xrightarrow{dil. HCl}$
$\quad / \ \backslash$
$CH_3 \quad CH_3$

$CH_3 \quad OH$
$\quad \backslash \ /$
$\qquad C \qquad + Mg^{\oplus\oplus} + Br^\ominus + Cl^\ominus$
$\quad / \ \backslash$
$CH_3 \quad CH_3$

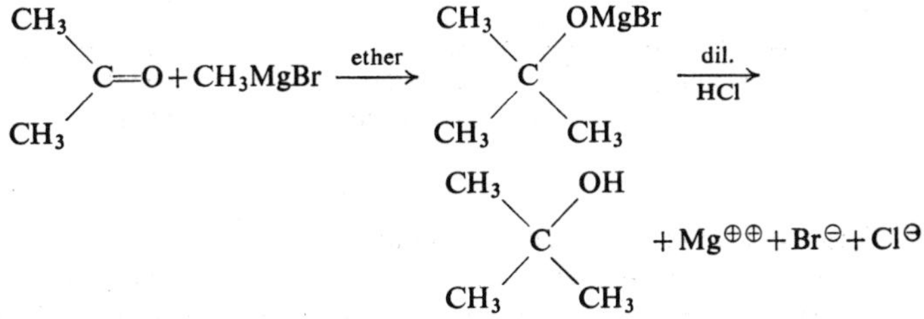

3 (a) $C_6H_5-CH(CH_3)_2$ (b) As (a). (c) $C_6H_5-C(CH_3)_2-CH(CH_3)_2$

4 The secondary carbonium ion formed, $(CH_3)_3C.\overset{\oplus}{C}H.CH_3$ rearranges to the more stable tertiary carbonium ion $(CH_3)_2\overset{\oplus}{C}.CH(CH_3)_2$. This then deprotonates to give tetramethyl ethylene, $(CH_3)_2C{=}C(CH_3)_2$, as the major product.

5.5 page 100

1

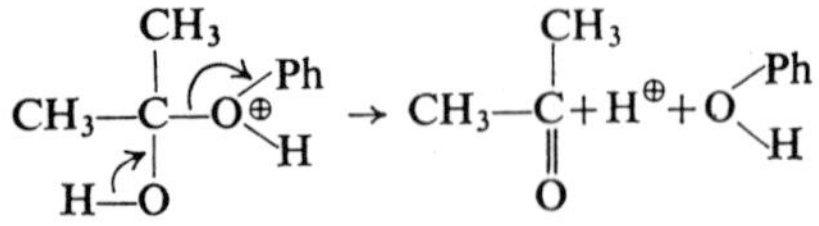

2

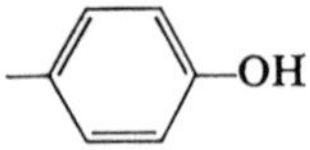

migrates because, of the three possible groups, it is the best electron donor (—OH is activating). Products are

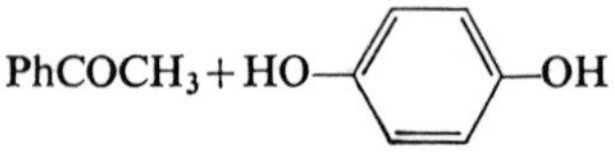

Miscellaneous exercises p. 102

1 (a) CH_3CHO (b) CH_3COCH_3
   Markovnikov's rule predicts product (b).
2 (a) C_6H_5—Cl   (b) C_6H_5—CN
3 The phenyl group, C_6H_5, is a more effective electron donor than methyl, hence the carbonyl carbon atom of benzaldehyde is less partially positive than that of acetaldehyde.

4

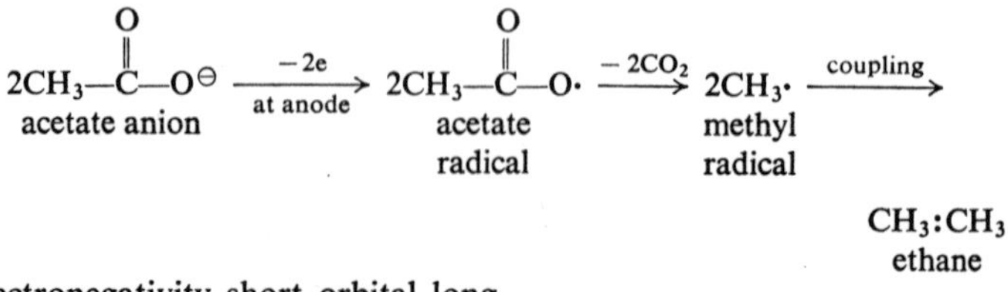

5 Electronegativity, short, orbital, long.
6 The Cl atom is partially drawn into the π electrons of the olefinic bond (p orbital overlap). This tends to decrease slightly the length of the C—Cl bond and to increase the length of the olefinic bond. The partial double

bond character of the C—Cl bond results in the Cl atom being more strongly held, and hence less reactive, than in a normal alkyl chloride.

7

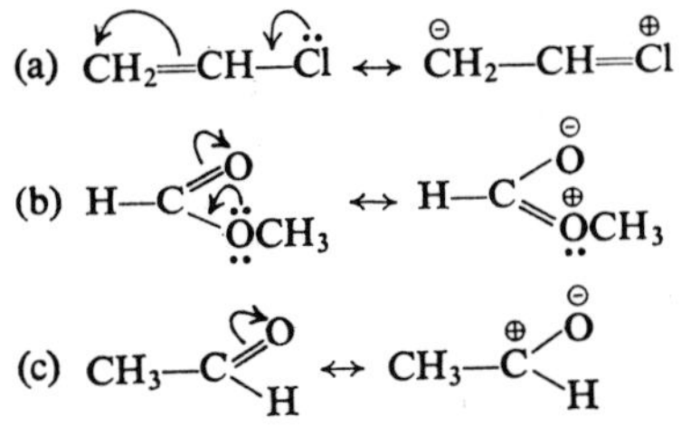

(a) $CH_2\!\!=\!\!CH\!\!-\!\!\ddot{C}l \leftrightarrow \overset{\ominus}{C}H_2\!\!-\!\!CH\!\!=\!\!\overset{\oplus}{C}l$

(b)

(c)

8

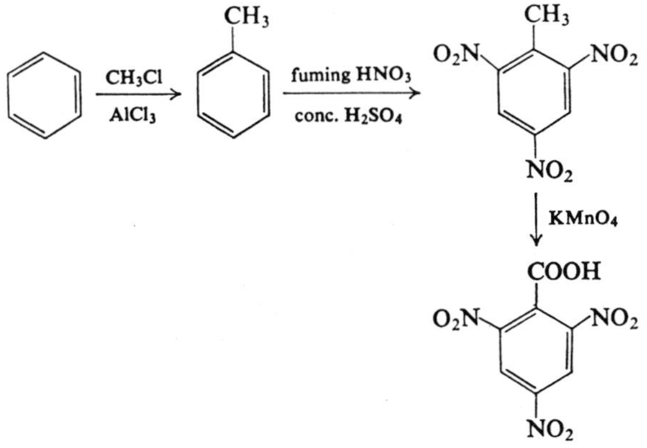

9 The two olefinic bonds of the 1,3 isomer can form a resonant system by a limited π orbital overlap. The double bonds of the 1,4 isomer are too far away from each other to do this. The 1,3 isomer therefore has a resonance energy of about 5 kcal/mole.

10 Electron attraction by the two carbonyl groups renders the C—H bonds of the central CH_2 group considerably weaker than those of a normal paraffin. Hence proton loss is likely to occur from CH_2 to give an anion which is significantly resonance stabilized (3 canonical forms). The compound is therefore a weak acid.

11

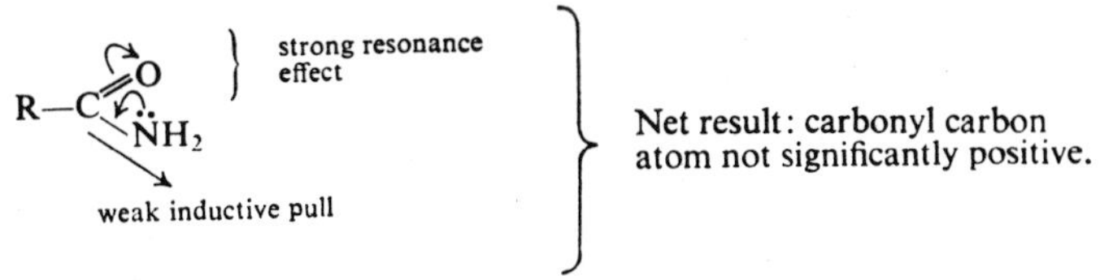

Net result: carbonyl carbon atom not significantly positive.

116

12 (a) $CH_2{=}CH_2$

(b) $CH_3.CH_2.CH{=}CH_2 + CH_3.CH{=}CH.CH_3$

(c) $(CH_3)_2C{=}CH.CH_3 + CH_3{-}\underset{\underset{\displaystyle CH_2}{\|}}{C}{-}CH_2.CH_3$

Glossary

Angström unit [Å]	10^{-8} cm.
Autoxidation	direct oxidation of a substance by molecular oxygen which does not involve combustion.
Bond energy	the energy required to rupture a chemical bond.
Canonical forms	the possible written chemical structures for a given chemical entity.
Carbanion	an organic anion with the negative charge on carbon.
Carbonium ion	an organic cation with the positive charge on carbon.
Chain reaction	a reaction which after being started, continues to propagate itself.
Cis	on the same side of.
C-nitrosation	substitution of the group —N=O at carbon.
Debye unit (D)	10^{-18} e.s.u. cm.
Decarboxylation	loss of carbon dioxide.
Dimerization	the joining of two like molecules (the monomer) to form one molecule (the dimer).
Dipolar	having two electrically opposite poles resulting from a separation of charge within a single entity.
Dipole moment	a measure of the electrical asymmetry of a dipolar molecule; = charge x distance of separation in the molecule; measured in Debye units (page 12).
Electronegativity	the ability of an atom to attract electrons towards itself.
Electronegativity coefficient	a measure of the relative electron attracting powers of different atoms within a covalent bond (page 9).
Electrophilic	electron seeking (page 5).
1-2 elimination	the removal of two atoms or groups from adjacent carbon atoms in a molecule.
Good leaving group	a stable entity which leaves the reactant molecule under consideration during a chemical reaction.
Heterolytic fission	rupture of a chemical bond with unequal sharing of the bond's electrons (page 3).
Homolytic fission	rupture of a chemical bond with equal sharing of the bond's electrons (page 3).
Inductive effect	the uneven sharing of the bonding electrons by the atoms which they join, resulting from differences in electronegativity of the two atoms.
Intermediate bond	a chemical bond which cannot be accurately classified as single, double, triple etc. due to resonance within the molecule.
Nucleophilic	nucleus seeking (page 5).
pK value	a measure of the strength of a weak acid or weak base = —$\log_{10}$ (dissociation constant).

117

118

Polarization	electrostatic distortion.
Radical	an electrically neutral species with an odd unpaired electron.
Resonance	the concept which chemists use to describe the actual structure of a chemical species in which some of the bonding electrons are delocalized so that there is more than one way of writing its valence bond structure.
Resonance energy	the difference between the actual energy content of a chemical species and that of its canonical forms.
Stereochemistry	the study of the properties of substances depending upon the arrangement of the atoms in three-dimensions.
Stereoisomers	compounds having the same molecular formula and with corresponding atoms linked to the same atoms, but having the atoms arranged differently in space; they can be either geometrical or optical isomers.
Substrate	the reactant molecule which combines with the reagent to yield the product in a chemical process.
Trans	on opposite sides of.

Select Bibliography

The following books and papers will be found to be stimulating for the reader requiring further information about organic reaction pathways.

AUSTIN, A. T. 'Nitrosation in Organic Chemistry', *Science Progress*. October 1961.

BARTLETT, P. D. 'Reaction Mechanisms', in *Perspectives in Organic Chemistry*, ed. A. R. Todd. Interscience, 1956.

CONROW, K. and MCDONALD, R. N. *Deductive Organic Chemistry*, Addison-Wesley, 1966.

GEISSMAN, T. A. *Principles of Organic Chemistry*, Freeman, 1962.

GOULD, E. S. *Mechanism and Structure in Organic Chemistry*, Holt-Rinehart & Winston, 1959.

HINE, J. *Physical Organic Chemistry*, McGraw-Hill, 1962.

INGOLD, C. K. *Introduction to Structure in Organic Chemistry*, Bell, 1956.

SYKES, P. *A Guidebook to Mechanism in Organic Chemistry*, Longmans, 1961.

TEDDER, J. M. and NECHVATAL, A. *Basic Organic Chemistry*, Wiley, 1966.

WATERS, W. A. *Mechanisms of Oxidation of Organic Compounds*, Methuen, 1964.

Index